THE
TRANSITS
OF
VENUS

HARRY WOOLF

ARNO PRESS

A New York Times Company

New York • 1981

Editorial Supervision: RITA LAWN

Reprint Edition 1981 by Arno Press Inc.
Copyright © 1959 by Princeton University Press
Reprinted by permission of Princeton University Press

THE DEVELOPMENT OF SCIENCE: Sources for the History of Science
ISBN for complete set: 0-405-13850-4
See last pages of this volume for titles.

Manufactured in the United States of America

Library of Congress Cataloging in Publication Data

Woolf, Harry.
    The transits of Venus.

    (The Development of science)
    Reprint of the ed. published by Princeton University
Press, Princeton, N. J.
    Bibliography: p.
    Includes index.
    1. Venus (Planet), Transit of. 2. Astronomy--
History. 3. Science--History. I. Title. II. Se-
ries: Development of science.
[QB509.W75 1981]   523.9'6   80-2150
ISBN 0-405-13959-4

THE TRANSITS OF VENUS
A STUDY OF EIGHTEENTH-CENTURY SCIENCE

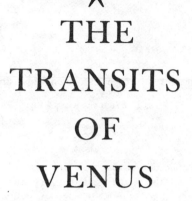

THE
TRANSITS
OF
VENUS

A STUDY OF
EIGHTEENTH-CENTURY
SCIENCE

BY HARRY WOOLF

PRINCETON, NEW JERSEY
PRINCETON UNIVERSITY PRESS
1959

TO MY PARENTS

A SENSE OF JOY

THE *Transits of Venus* deals with the eighteenth-century attempt to complete the Newtonian system of the world by determining its actual scalar dimensions. In the fortunate occurrence of the rare transits of Venus, the scientific mind of the Enlightenment believed it had discovered the mechanism for obtaining one of the basic natural constants of physical astronomy, the solar parallax. With this invariant definitively determined, after the great Newtonian achievement that established the laws of its dynamic behavior, the size of the solar system could be calculated accurately, and the "final" problem of astronomy solved.

The usefulness of the transits of Venus in determining the fundamental unit of astronomical measurement, the solar parallax, or the mean distance of the earth from the sun, was first demonstrated by Edmund Halley in a series of papers published between the closing years of the seventeenth century and the second decade of the eighteenth. Essentially, Halley's procedure consisted of observing the moments of inner contact between Venus and the sun during the planet's ingress and egress from the solar disk. Observations of this phenomenon from different parts of the globe, he reasoned, would lead to differences in the duration of time which the planet appeared to spend upon the sun, and from this information the solar parallax could be accurately determined. In the years following Halley's death, the transit method was modified by Joseph-Nicolas Delisle so that observations of contact at either ingress or egress would be adequate. Delisle's world-wide correspondence also helped to stimulate interest in the transits everywhere, and through his letters it is possible to take some measure of the breadth and depth of interest which the scientific community maintained in the transits during the years preceding their occurrence.

Primarily because of the efforts of the Académie royale des sciences and the Royal Society of London, but also through the assistance of the crowned heads of Europe and the provincial

vii

assemblies of America, astronomical expeditions were dispatched to all parts of the world to observe the transits in 1761 and 1769. With skilled observers selected from the substantial reservoir of scientific talent available by the middle of the eighteenth century, and equipped with precise astronomical instruments drawn from the burgeoning industry in scientific instruments of the same period, the expeditions were the best of their kind to date.

The calculations derived from the transit observations of 1761 considerably reduced the element of uncertainty in the solar parallax determined by pre-transit techniques. But the general result remained unsatisfactory for what was supposed to be the precise determination of a natural constant. However, the fortunate occurrence of transits of Venus in pairs, eight years apart, presented the scientific community with a second chance which it could not neglect. The transit of 1769 was thus observed with the greatest care, and the superior conclusions drawn from its observations were a measure of both the value of the experience of 1761 and the technical improvements which had been made in astronomy during the decade.

On the whole, this book is the story of the eighteenth-century attempt to determine the dimensions of the solar system, that concern which brought more interests to a single focus than any other scientific problem in the Age of Reason. It is based primarily upon the archival record of scientific expeditions, the unpublished deliberations of learned academies and their national governments, and the private correspondence of scientists as well as the printed results of their work. Thus, resting upon the traditional sources of history, the *Transits of Venus* is an attempt to set down the narrative, where none has existed before, of those important events in the history of science that bore witness to the establishment of the modern international community in science. Indeed, the rare transits of Venus and the kind of attention which they received in the eighteenth century have provided a sterling opportunity for analyzing the organization and practice of science during the Enlightenment.

My thanks are due to many whose names can never be recorded in this traditional statement of gratitude and respect.

There are those whose lives have touched mine with such understanding and sympathy, with rewards in education, encouragement, and friendship so great as to have given me purpose and direction after troubled times. Thanks for so much freely given can never be enough, save the reminder that "I have promises to keep/And miles to go before I sleep."

Special thanks are due to Professor Henry Guerlac, who first suggested the subject and encouraged me to explore its possibilities in his Seminar in the History of Science at Cornell University. Mr. A. E. Lownes of Providence, Rhode Island, was kind enough to lend me documents and diagrams from his Transit of Venus collection. The award of a Fulbright grant made it possible for me to undertake the foreign research germane to this study, and I would here like to express my deepest appreciation for that support. An additional grant from the Research Committee of the University of Washington covered the cost of the final typing, so well accomplished by Nora Giesey.

I am especially grateful to all those who helped me overcome the initial difficulties of working abroad by friendly advice and innumerable kindnesses. In particular I wish to thank M. Denizet of the Service historique de la marine, M. Arbey of the Observatoire de Paris, M. Dupuy of the Institut géographique, and Mlle Boy of the Bibliothèque Ste. Geneviève. For permission to work in the archives of the Royal Society, the Académie des sciences and the Observatoire de Paris, I am grateful to the Officers of the Royal Society, to the *Secrétaires perpétuels* of the Académie des sciences, and to M. André Danjon, *Directeur* of the Observatoire de Paris.

For the right to reproduce certain texts and illustrative material, I wish to express my thanks to the following individuals and institutions for permission to publish from their holdings:

The lines "But I have promises to keep/And miles to go before I sleep" are from Robert Frost's "Stopping by Woods on a Snowy Evening" in *Collected Poems of Robert Frost*, New York: Henry Holt & Co., 1942.

The photograph of the bust of Alexandre Gui Pingré was obtained from Monsieur Noel Le Boyer of Paris, and the permission to publish it together with the various texts, diagrams and other Pingré material extracted from the Archives of the Bibliothèque Sainte-Geneviève, is the generous grant of Monsieur G. de Valous, the *Conservateur en chef* of that excellent library.

The photograph of the Gregorian telescope by Benjamin Martin is reproduced with the permission of the Superintendent, Science Museum, South Kensington, London, "Crown Copyright, Science Museum, London."

Professor I. Bernard Cohen of Harvard University has supplied me with the superb portrait of Father Maximilian Hell, and Professor Charles H. Smiley of Brown University has done the same for the photograph of the Joseph Brown-Benjamin West telescope.

Chapter III, in a somewhat different form and title, appeared in the *William and Mary Quarterly* for October 1956. The Institute of Early American History and Culture, publishers of that journal, have kindly consented to the reprinting of most of the article.

For guiding me through the difficulties of the Latin sources, my warmest thanks go to Mrs. Anne L. Spitzer and to my colleague Dr. Paul Dietrichson of the Department of Philosophy for his help with the Swedish material. For generous assistance in the preparation of the index, I wish to express my thanks to Misses Phyllis Brooks and Dorothy Stratton. Under great difficulties, my sister Paula typed the first draft of this manuscript. That our friendship and affection have survived is a fitting measure of her skill and devotion. To my wife and children, whose patience and tolerance were tested to the extreme, no thanks can ever be sufficient, no appreciation deep enough.

Seattle, Washington   HARRY WOOLF
June 1958

CONTENTS

TEXT DIAGRAMS

xiii

THE TRANSITS OF VENUS

A STUDY OF EIGHTEENTH-CENTURY
SCIENCE

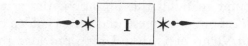

THE ASTRONOMICAL UNIT AND THE
TRANSITS OF VENUS

In the history of modern science, great care has been lavished upon certain basic constants. Atomic weights, the velocity of light, and the constant of gravitation, to name but a few, serve as keys to the closed doors of nature, and their determination represents a cumulative series of human triumphs, a measure of constancy in a world of change. The historical reconstruction of the eighteenth-century attempt to calculate one of the natural constants of physical astronomy, the sun's mean distance from the earth, constitutes the main purpose of this study.

The importance of this measurement can hardly be exaggerated. Expressed as the mean radius of the earth's orbit, it becomes the standard measure for the universe, a kind of celestial meter stick of no less importance than its terrestrial counterpart. Kepler's laws of planetary motion and Newton's mechanics involved dynamic considerations which were independent of the actual dimensions of the system for which they were established. It was sufficient to know the relative position of one planet with respect to another. The adoption of some arbitrary value for the distance between the earth and the sun would make possible, according to Kepler's third law,[1] the determination of the relative disposition of the other members of the solar system. But this is comparable to expressing the metric distance between New York and Philadelphia without knowing the length of the meter. A major scientific aspiration of the day was, therefore, that of solving the problem of the actual scalar dimensions of the solar system and thereby completing this fundamental gap in the Newtonian system of the world.

[1] The squares of the times of revolution (periods) of the planets are proportional to the cubes of their mean distances from the sun.

3

In the attempt to fulfill this ambition, the astronomers of the eighteenth century were presented with a golden opportunity. The transits of Venus of 1761 and 1769 provided the necessary celestial mechanism for determining the most accurate calculation of the solar distance obtainable down to the middle of the nineteenth century. These infrequent transits of Venus across the face of the sun were actually unobserved until the seventeenth century. As a matter of fact, only five such events have been observed in the history of astronomy: those of 1639, 1761 and 1769, and 1874 and 1882. We shall have occasion to refer to them again. For the moment let us note that these events have a regular cycle of 243 years, made up of four irregular intervals which are alternately short and long: the short ones are always eight years apart and the long ones alternately 121½ and 105½ years.[2] The interesting phenomenon that transits occur in pairs at intervals of eight years, when they do occur at all, arises from the fact that the time for eight revolutions of the earth is equal to thirteen for Venus.

Besides the astronomical opportunity of observing a relatively rare phenomenon provided by the eighteenth-century transits, their occurrence is significant for additional reasons. They took place at the time of the most fruitful union of mathematics and observational astronomy that had yet occurred and, equally important, at a time when the principles of the Enlightenment stood at high noon. The former suggests that the scientific means for utilizing the transits would be at hand, and the latter implies that enough interest would be present to encourage and finance the necessary research. This is exactly what happened; the conjunction of enlightened interest and scientific practice, actually achieved in the observations of the transits, also gave rise to the first international, cooperative scientific expeditions in modern history.

II

The problem of determining the actual scale of the astronomical world has a history of its own, and did not spring full-blown

[2] Simon Newcomb, *Popular Astronomy* (New York: Harper & Bros., 1878), p. 177.

from the mind of eighteenth-century man. Before any methods of applying a transit of Venus to the solution of the problem can be discussed, a brief sketch of its evolution seems in order. The first evaluation of the solar distance in the Western tradition is derived from the school of Miletus in the sixth century B.C. Anaximander, who speculated on both the distances of the sun and the moon from the earth, put the former at a distance of twenty-seven times the diameter of the earth and the latter at nineteen times the same unit.[3] There seems to have been no distinctive geometric calculation behind these figures, in spite of several attempts by later scholars to credit Anaximander with such a technique.[4] A more likely explanation appears to be that these values were arrived at through some sort of number mysticism.[5]

An illustration of the same kind of approach, so well known that it need only be mentioned rather than elaborated, is to be found in the Pythagorean treatment of the problem. This concerns the harmony of the spheres as a device descriptive of the proportional relationship between the heavenly bodies. The ratios between the distances of these bodies corresponded to successive notes in the Greek musical scale, but how they were actually translated into estimated relative distances, if they were at all by the early Pythagoreans remains unknown.[6] In any case, we learn of

[3] T. L. Heath, *Aristarchus of Samos* (Oxford: Clarendon, 1913), p. 37. J. L. E. Dreyer, in *A History of Astronomy from Thales to Kepler* (2nd edn.; New York: Dover Publications, 1953), p. 178, mentions an alternative interpretation of the fragment of Anaximander upon which this is based, that is, Aetius 2, 24: 354. This can be most readily found in M. C. Nahm, *Selections from Early Greek Philosophy* (3rd edn.; New York: Appleton-Century-Crofts, 1947), pp. 64-65. P. Tannery presents still another interpretation in *Pour l'histoire de la science Hellène* (2nd edn.; Paris: Gauthier-Villars et Cie., 1930), pp. 94, 121. This could possibly mean that the solar distance is twenty-seven times greater than that of the moon, but the best evaluation of the fragment seems to be Heath's. Incidentally, Nahm's translation simply says that "the circle of the sun is twenty-eight times as large as the earth," but Tannery refers to the earth's diameter as the basis for comparison.

[4] The interpretations of Teichmüller (*Studien zur Geschichte der Begriffe*), Neuhäuser (*Anaximander Milesius*), Sartorius (*Die Entwicklung der Astronomie bei den Griechen bis Anaxagoras und Empedokles*), Tannery (*Pour l'histoire de la science Hellène*), and Zeller (*Philosophie der Griechen*) are discussed and rejected in this respect by Heath.

[5] Heath, *Aristarchus of Samos*, p. 38.

[6] *ibid.*, p. 111.

5

an approximation of the solar distance derived from Pythagorean influence. According to the *Ars Eudoxi*,[7] Eudoxus is supposed to have arrived at a relative distance between the sun and the moon of nine to one.[8] He reasoned that the sun was nine times larger than the moon and that because they both appeared to be the same size, it followed that the distance of the sun was nine times greater than the distance of the moon. The Eudoxian method persisted through the Middle Ages, but though it contributed little to the ultimate solution of the problem, "it deserves honourable mention in the history of human progress."[9]

The question whether Aristarchus of Samos had any predecessors in the attempt to calculate the solar distance mathematically remains a highly speculative one. It is known that from about the middle of the fourth century B.C., numerous studies were made of the sizes and distances of the planets.[10] Indeed, the name of one author of such treatises is known, Philip of Opus, a disciple of Plato; but of books relating to our subject we have only the titles: *On the Distance of the Sun; On the Size of the Sun and the Earth*; and *On the Eclipse of the Moon.*[11] Nothing is known of their actual contents, and it is to Aristarchus that we must grant first place for a calculation of the solar distance which, for all of its weakness, may be termed scientific.

The single work by which we know Aristarchus, *On the Sizes and Distances of the Sun and Moon*,[12] deserves discussion here

[7] *ibid.*, p. 112. This so-called papyrus of Eudoxus (408-355 B.C.) was written between 193 and 165 B.C. I am aware that both the late date of the composition, some 300 years after Pythagoras and about 180 years after Eudoxus, and the fact that in citing Eudoxus I am not even referring to a pure Pythagorean at all result in a rather weak case for a Pythagorean statement on the solar distance. Yet I do feel that one can be made and that this is as good a choice as any.

[8] Dreyer, *op.cit.*, p. 180. See also T. L. Heath, *op.cit.*, p. 196, and T. L. Heath, (ed.), *The Works of Archimedes* (Cambridge: The University Press, 1897), p. 223.

[9] *ibid.*, p. 181.

[10] F. Enriques and G. De Santillana, *Mathématiques et astronomie de la période Hellénique* (Paris: Hermann & Cie., 1939), p. 38.

[11] Heath, *Aristarchus of Samos*, p. 320. We are told that he was the editor of the *Laws* of Plato and the author of the *Epinomis*.

[12] Text and English translation in Heath, *Aristarchus of Samos*. The first Latin translation of this text by G. Valla was printed in Venice, 1498. In the nineteenth century it was translated into French by Fortia d'Urban as a part of his *Histoire d'Aristarque de Samos* (Paris: n.p., 1823). The work itself can be loosely dated by the use of observations of the summer solstice of 280 B.C. which Aristarchus made. See Enriques and Santillana, *op.cit.*, p. 39.

because of its historical position and the neat simplicity of its method. Aristarchus assumed that the moon receives its light from the sun. He then reasoned that when the moon is half full, the sun must be directly in line with its bright side. A line drawn from an earth-bound observer to the inner edge of the darkened half-portion of the moon would therefore be at right angles to the straight line joining the sun and the moon. A third line could then be drawn from the observer's position to the sun to form a triangle between earth, moon, and sun.

Aristarchus was able to measure one of the angles of this right triangle, and thereby obtain a ratio between the distance from the earth to the sun and the distance from the earth to the moon. The distance from the earth to the sun proved to be eighteen to twenty times greater than the distance from the earth to the moon.[13] T. L. Heath indicates the actual solar distance which Aristarchus arrived at to be 180 times the earth's mean diameter.[14] This is about 1,400,000 miles, a value which makes Aristarchus' solar system rather small by modern estimates.[15] But as Gino Loria writes, ". . . bien que cet écrit, [Sur les Grandeurs et les Distances du Soleil et de la Lune] à cause de son imprécision dans les résultats numériques, ne puisse plus obtenir une place dans la bibliothèque de travail d'un astronome, il est toutefois, par son orientation nettement géométrique, bien

[13] Enriques and Santillana, op.cit., p. 40.

[14] Heath, Aristarchus of Samos, p. 350.

[15] An interesting series of errors seems to occur at this point if one wishes to determine the actual distance in miles. Heath gives a value of 1,716 geographical miles for the earth's mean diameter (Heath, Aristarchus of Samos, p. 350). This is obviously out of the question. The mean radius of the earth is 3,958.82 miles (F. R. Moulton, Astronomy [New York: The Macmillan Co., 1931], p. 86), and by no definition of the geographical or nautical mile can these two figures be equated. On the other hand, the error is confounded by Enriques and Santillana, op.cit., pp. 40-45. "[P]our mieux résumer les informations du présent chapitre" (p. 45), they have borrowed Heath's figure (for Aristarchus) of the solar distance being 180 times the earth's diameter. Yet only five pages earlier (p. 40), they write that "il [Aristarchus] fut amené à estimer la distance du Soleil égale à environ 1.400 rayons terrestres." This value of 1,400 radii for the solar distance is undocumented, and clearly bears no relation to the figure 180 cited by Heath. Granted that Heath is speaking in terms of the earth's diameter, if his value is correct, then the solar distance expressed in terms of earth's radius ought simply to be twice 180 or 360, and not 1,400. For an actual evaluation then, the solar distance is 360 times the earth's mean radius, or 360 x ca. 3900 miles = 1,404,000 miles. The solar distance is actually more than 60 times as great.

digne d'être mentionné à côté des autres ouvrages parus pendant l'âge d'or de la . . . [science] grecque."[16]

In discussing the evolution of specific values for the solar distance, it is equally important to note when the notion of parallax itself (to be defined below.) came into being. Although the evidence is far from satisfactory, the recognition of this extremely useful concept in astronomy seems due to Archimedes, at least with respect to the parallax of the sun.[17] A more detailed explanation of the solar parallax and its measurement will be found below, but for the sake of immediate discussion, the solar parallax may be defined as the angular size (i.e. in terms of the radius or diameter) of the earth as seen from the sun. Impressed by the greatly expanded dimensions of the universe imagined by astronomers, Archimedes sought to demonstrate, in *The Sand-Reckoner*, that this growth offered no limitations to the power of mathematics to delineate the universe arithmetically.[18] In traditional discussions of the size and distance of the sun and moon, such as in the famous treatise by Aristarchus, it was generally assumed that the earth was a point in relation to the sphere in which the sun moved. This assumption automatically precluded the possibility of any consideration of solar parallax, for with the earth considered as a point, it could not provide any dimension, in chord or arc-length, to subtend an angle at the sun. But in his discussion of the apparent angular diameter of the sun that forms a part of *The Sand-Reckoner*, Archimedes abandoned this point of view. Its details need not concern us here, save to note that Archimedes is "recognizing parallax in the case of the sun, apparently for the first time."[19]

Three significant evaluations of the solar distance remain to be considered. Though widely differing among themselves, they nevertheless represent the best astronomy of antiquity on this

[16] *Histoire des sciences mathématiques dans l'antiquité Hellênique* (Paris: Gauthier-Villars et Cie., 1929), p. 168.

[17] Heath, *Aristarchus of Samos*, p. 348.

[18] In his own words, "A sphere of the size attributed by Aristarchus to the sphere of the fixed stars would contain a number of grains of sand less than 10,000,000 units of the eighth order of numbers (or 10^{56} plus $7 = 10^{63}$)." Archimedes, "The Sand-Reckoner," *The Works of Archimedes*, ed. T. L. Heath, p. 232.

[19] Heath, *Aristarchus of Samos*, p. 348.

question. The closest to the truth and boldest of all the estimates down to modern times was that of Posidonius,[20] who placed the sun at a distance of 6,545 times the earth's diameter, or about 51,700,000 miles.[21] The earlier parts of his method seem to stem from Archimedes' treatise on large numbers,[22] but the details of his actual work are not directly known. Hipparchus established the solar distance as 1,245 times the earth's diameter; Ptolemy, who owed so much to this greatest of ancient astronomers, apparently canceled his indebtedness at this point for he not only fails to mention this figure but goes on to indicate a value of his own, namely, 605 terrestrial diameters.[23] It is rather striking, in view of the connection between the two, that the more accurate determination of Hipparchus should remain unknown, while Ptolemy's figure, derived from a combination of the work of Aristarchus and his own calculation of the lunar parallax and expressed in angular measure as 3 minutes,[24] should persist down to early modern times.

III

Toward the end of the century which first called into question the principles of the entire Ptolemaic system, contemporary opinion on the solar distance was equally challenged. The authoritarian status which Ptolemy's writings acquired during the Middle Ages helped to guarantee the acceptance of the three-minute

[20] A Stoic philosopher and polymath who was born in Apamea in Syria ca. 135 B.C. and died in Rome ca. 51 B.C. His interests included philosophy, astronomy, geography, meteorology, and history. His voluminous publications exercised a continuous influence on Romans like Cicero, Lucretius, and Seneca, and on later medieval thought. Few of his actual writings survive, however. M. R. Cohen and I. E. Drabkin, A Source Book in Greek Science (New York: McGraw-Hill Book Co., 1948), p. 90.

[21] Heath, Aristarchus of Samos, p. 350. From Hultsch's table of sizes and distances for the sun and moon by ancient Greek calculators, the mean distance of the earth from the sun—in terms of the earth's diameter—is given as 6,545 for Posidonius. Taking the earth's diameter as ca. 7,900 miles, one may then calculate a solar distance of 6,545 x 7,900 = 51,705,500 miles.

[22] ibid., pp. 344-349.

[23] Claudius Ptolemy, The Almagest, trans. R. C. Taliaferro ("Great Books of the Western World," Vol. 16 [Chicago: Encyclopaedia Britannica, Inc., 1952]), pp. 173-175.

[24] R. Grant, History of Physical Astronomy from the Earliest Ages to the Middle of the Nineteenth Century (London: H. G. Bohn, 1852), p. 211.

solar parallax to the time of Tycho Brahe. Even Copernicus differed only slightly from Ptolemy, setting the distance at 750 times the earth's diameter as compared to Ptolemy's 650.[25] Tycho made no attempt to re-evaluate the solar distance, but Kepler, the brilliant co-worker of his last years, took it upon himself to re-examine the problem. Making use of the accumulated and very reliable observations of his master, Kepler concluded that Ptolemy's estimate of the solar parallax was much too large and that, in fact, the parallax could not be greater than one minute, and was very likely much below that upper limit.[26] In 1627 Kepler completed the Rudolphine Tables, and with these as a basis he was able to predict with considerable accuracy the motions of the inferior planets. A precise knowledge of the motions of these inner planets, he reasoned, would make it possible to decide just when they would pass across the face of the sun and be visible from the earth during the passage; that is to say, just when they would transit. Two years later he published a pamphlet in which he predicted a transit of Mercury for 7 November and one of Venus for 6 December 1631, though without connecting the latter to the question of the solar distance.[27]

Pierre Gassendi observed the transit of Mercury of 1631 at Paris, thereby becoming one of the first individuals ever to witness the transit of a planet. It was once assumed that he was alone in this observation, but at least three others are now known to have seen Mercury on the solar disk during the same transit.[28] Nevertheless, Gassendi's technique is worth recording. Making use of the principle of the camera obscura, he admitted the solar light into a darkened room through a small opening in the window. The image of the sun was then received upon a

[25] Heath, *Aristarchus of Samos*, p. 343.

[26] Grant, *op.cit.*, p. 211.

[27] *ibid.*, p. 415. The pamphlet bore the Latin title: "Admonitio ad Astronomos rerumque celestium studiosos, de miris rarisque anni 1631 phaenomenis, Veneris puta et Mercurii in solem incursu," ([Leipzig]: 1629).

[28] P. Humbert, *L'Oeuvre astronomique de Gassendi* (Paris: Hermann & Cie., 1936), p. 21. They were Remus Quietanus at Rouffach, a Father Cysatus at Innsbruck, and an anonymous Jesuit at Ingolstadt. The observations of these men lacked any kind of astronomical precision, hence they were without scientific value.

white screen, and a circle drawn corresponding to its exact outline. Axes were constructed to divide the circular area into four equal parts, and the circle itself was divided into 360 degrees. To obtain the time at the moment of observation, Gassendi placed an assistant in the room above him, whose duty it was to observe the altitude of the sun with a 2-foot quadrant whenever Gassendi signaled by stamping his foot.[29]

When the transit took place, Gassendi was, for the most part, able to track its path across the sun by drawing a chord corresponding to the movement of Mercury's image across his screen. Furthermore, he was able to estimate its apparent diameter (about 20 seconds in angular measure) and, by reducing Kepler's calculations to the Paris meridian, to determine the observed time of the transit.[30] By all counts it was therefore a successful scientific observation. With the publication of his results, even his sceptical contemporaries were convinced of his achievement. His good friend and occasional collaborator, Peiresc,[31] wrote him to say that "cette belle observation que vous avez faite du passage et sortie de Mercure devant la face du Soleil . . . j'estime l'une des plus dignes qui se soit faicte de beaucoup de siècles. . . ."[32]

Gassendi attempted to repeat his success with the Mercury transit for that of Venus, but an error in the Rudolphine Tables prevented him from doing so, the planet itself crossing the face of the sun during the European night between the sixth and

[29] ibid., p. 22. See also Grant, op.cit., pp. 415-416 and Gassendi, "Mercuris in Sole visus, & Venus invisa Parisiis anno 1631 pro Voto & admonitione Keppleri: cum observatus quibusdam aliis," Opera Omnia (Lyon: Chez Laurent Anison, 1658), IV, 537ff. The circumstances following this memorable observation were also reported by Gassendi in a letter to his friend Wilhelm Schickhard, Professor of Mathematics and Hebrew at the University of Tübingen.

[30] Grant, op.cit., p. 416. He noted that the actual time of the transit was 4 hours 49 minutes and 30 seconds in advance of the computed time.

[31] Nicolas-Claude Fabri de Peiresc (1580-1637) was one of the outstanding men of his day. A member of the Parlement of Aix-en-Provence, he was a geographer, astronomer, archeologist, botanist, and physiologist. In 1610 he discovered the nebula in Orion and he frequently collaborated with Gassendi on astronomical observations. Cf. P. Humbert, Un Amateur: Peiresc, 1580-1637 (Paris: Desclée de Brouwer, 1933).

[32] From an unpublished letter by Peiresc, quoted in Humbert, L'Oeuvre astronomique de Gassendi, p. 23. In the same letter, Peiresc gives the details of his own fruitless attempt to observe the transit at Belgentier.

seventh of December. Consequently, no observation of the 1631 transit of Venus was made;[33] and, but for the early genius of Jeremiah Horrox (1619-1641), none would have been made of the 1639 transit either.

Kepler's tables had led to the prediction that the next transit of Venus after that of 1631 would take place in 1761, and that in 1639 the planet would be too far south to pass before the sun. The tables of Lansberg, on the other hand, had predicted that the planet would pass over the northern tip of the sun.[34] Experience with the tables of Lansberg, however, at least since 1636,[35] had taught Jeremiah Horrox to doubt their value. The Rudolphine Tables, while infinitely superior in the hypotheses upon which they were founded, also contained some imperfections. Accordingly, well before he faced the transit problem, Horrox was engaged in the double process of emending Kepler's tables and refuting those of Lansberg.[36] It was thus while reconsidering the more basic errors in the Lansberg tables that he discovered that the transit of Venus would indeed take place in 1639. "Whilst thus engaged," he writes, "I received my first intimation of this remarkable conjunction of Venus with the Sun; and I regarded it as a very fortunate occurrence, inasmuch as about the beginning of October, 1639, it induced me, in expectation of so grand a spectacle, to observe with increased attention. I pardon, in the meantime, the miserable arrogance of the Belgian astronomer, who has overloaded his useless tables with such unmerited praise . . . deeming it a sufficient reward that I was thereby led to consider and forsee the appearance of Venus in the Sun."[37]

[33] Grant, op.cit., p. 419.

[34] ibid., p. 420.

[35] A. B. Whatton, The Transit of Venus across the Sun . . . by the Rev. Jeremiah Horrox . . . to which is prefixed A Memoir of his Life and Labours (London: W. Macintosh, 1859), p. 18. In 1636, Horrox made the acquaintance of William Crabtree, a Manchester draper, who was also interested in astronomy. Correspondence between them on the subject of common observations and calculations taught them to recognize the imperfections in the Lansberg tables.

[36] ibid., pp. 22-25. By 1637 Horrox had prepared a treatise against Lansberg entitled, "Astronomiae Lansbergianae censura et cum Kepleriana comparatio," and shortly thereafter he produced another against Hortensius (a follower of Lansberg), who had attempted to depreciate the work of Tycho Brahe.

[37] Whatton, op.cit., pp. 110-111.

Horrox observed the transit of Venus of 1639, using much the same technique which Gassendi had employed for the transit of Mercury eight years earlier. However, instead of simply admitting the solar light into the darkened room through an aperture in the shutter, he substituted a telescope at the opening. In this way he could bring the image of the planet to a sharp focus on the screen.[38] His observation was a complete success. Save for his friend Crabtree, who had only occasionally glimpsed the planet on the sun, no other person had even witnessed the transit.

From this observation, Horrox was able to correct several elements in the theory of the planet's motion, and to state unequivocally that the diameter of Venus could not be greater than one minute. Though he was really unprepared to calculate the solar parallax with any degree of precision, he estimated that it could not exceed 14 seconds.[39] "I had intended to offer a more extended treatise on the Sun's parallax," he tells us in the conclusion to the Venus paper, but this work was never completed.[40] Nevertheless it was a bold conclusion, in view of Kepler's one-minute evaluation for the solar parallax; but the results of the eighteenth-century transit of Venus observations were to prove him closer to the truth than Kepler.

Though isolated and unsung in his own day, Horrox' early death was followed by substantial if only occasional praise. Newton praised him most highly, referring to his hypothesis on the inequalities in the moon's motion as the "most ingenious, and . . . of all, the most accurate,"[41] and crediting him with being the first to describe the elliptical path of the moon's motion around the earth.[42] Besides the work on the transit of Venus and the moon's orbit, Horrox had demonstrated his promise in a variety of projects: the detection of the inequality in the mean motion of Saturn and Jupiter, an essay on the nature and movement of comets, and an incomplete study on tides.[43] It is an impressive list for a man of twenty-two.

[38] ibid., pp. 44-45.   [39] ibid., p. 212.   [40] ibid., p. 215.
[41] F. Cajori, (ed.), Sir Isaac Newton's Mathematical Principles of Natural Philosophy and his System of the World (Berkeley: University of California Press, 1946), p. 577.
[42] ibid., p. 475.   [43] Whatton, op.cit., pp. 90-91.

Some twenty years rolled by after the work of Horrox before the first scientific calculation of the sun's distance was obtained, and this calculation was accomplished without a transit of Venus. Under the direction of Jean Dominique Cassini, in concert with Jean Richer at Cayenne,[44] simultaneous observations of the parallax of Mars were undertaken while the planet stood in opposition.[45] From the results thereby obtained, Cassini deduced a solar parallax of 9.5 seconds, or a distance of 87,000,000 miles. But Flamsteed, who had made independent observations of the same phenomenon, came to the conclusion that the parallax was 10 seconds, or a distance of 81,700,000 miles; while Picard, who had participated in Cassini's efforts in France, issued his own evaluation increasing the parallax to 20 seconds, a solar distance of 41,000,000 miles or about half of Flamsteed's value.[46] Thus, in spite of the employment of sounder scientific principles than had hitherto been used and the introduction of three new astronomical instruments that were to revolutionize observational astronomy—the pendulum clock, the filar micrometer, and the application of telescopes to the sights of graduated circles[47]—the range of uncertainty in determining the solar parallax remained enormous indeed.[48]

The suggestion that transit observations might prove useful in determining the solar parallax—a suggestion to which Horrox had vaguely given voice—remained only an implication until the work of James Gregory. This great Scots mathematician, who in his day "was held to be second only to Newton,"[49] unknowingly gave explicit expression to his English predecessor's idea. Some thirty-one propositions of his *Optica Promota* of

[44] J. W. Olmsted, "The Scientific Expedition of Jean Richer to Cayenne (1672-1673)," *Isis*, xxxiv (1942), 121-123.

[45] For an explanation of this term and others used in ordinary planetary discussion, see Appendix 1.

[46] Note that Flamsteed's evaluation is less accurate than the one attributed to Posidonius some 1800 years earlier. *Supra* p. 9.

[47] Olmsted, *op.cit.*, p. 123 and A. Wolf, *A History of Science, Technology and Philosophy in the XVIth and XVIIth Centuries* (2nd edn.; London: George Allen & Unwin Ltd., 1950), pp. 112-114, 165-174.

[48] C. F. Cassini, "Histoire abrégé de la parallaxe du Soleil," *Voyage en Californie pour l'observation du passage de Vénus sur le disque du Soleil. . . .* (Paris: A. Jombert, 1772), p. 122.

[49] H. W. Turnbull, *James Gregory, Tercentenary Memorial Volume* (London: G. Bell & Sons, 1939), p. vi.

1663 are devoted to astronomical problems; in a scholium to one of these propositions, in which he treats of the parallaxes of two planets in conjunction, he writes: "This problem has a very beautiful application, although perhaps laborious, in observations of Venus or Mercury when they obscure a small portion of the sun; for by means of such observations the parallax of the sun may be investigated."[50] Thus, in this marginal annotation to another problem, Gregory first stated what Horrox had vaguely implied. Theoretically, the transits of Mercury and Venus could indeed be used to determine the solar distance, but the development of a practical method for performing the task awaited the efforts of Edmund Halley.

Halley's first recorded interest in the possible utility of a transit of Venus for determining the solar parallax arose from his trip to St. Helena to observe the "stars about the south pole" and the transit of Mercury of 1677.[51] His first published reference to it, the result of that trip, was in his *Catalogus Stellarum Australium* of 1679.[52] He discussed the subject again in the *Philosophical Transactions* of the Royal Society for 1691, 1694, and 1716. "This sight [of a transit of Venus] which is by far the noblest astronomy affords," he announced in 1691, "is denied to mortals for a whole century, by the strict laws of motion. It will be afterwards shown, that by this observation alone, the distance of the sun, from the earth, might be determined. . . ."[53]

[50] J. Gregory, "Optica promota: seu abdita reflexorum et refractorum mysteria, geometricè enucleata; cui subnectitur appendix, subtilissimorum astronomiae problematon resolutionem exhibens" (London: J. Hayes, 1663), in F. Masères, *Scriptores Optici; or A Collection of Tracts Relating to Optics* (London: R. Wilks, 1823), pp. 99-101. The original Latin is as follows: "Hoc problema pulcherrimum habet usum, sed forsan laboriosum, in observationibus Veneris, vel Mercurii particulam Solis obscurantis: ex talibus enim solis parallaxis investigari poterit." I have used the translation by Grant, *op.cit.*, 428. Cf. W. T. Lynn, "Halley and the Transits of Venus," *The Observatory*, v, 62 (June 1882), 175.
[51] J. Ferguson, *Astronomy Explained Upon Sir Isaac Newton's Principles. . . .* (5th edn.; London: W. Strahan, 1772), p. 438.
[52] Grant, *op.cit.*, p. 429.
[53] E. Halley, "On the Visible Conjunctions of the Inferior Planets with the Sun," *Philosophical Transactions of the Royal Society of London from their Commencement, in 1665, to the Year 1800*, abridged and edited by C. Hutton, G. Shaw, and R. Pearson (London: C. & R. Baldwin, 1809), III (1683-1694), 454. I have used two editions of the *Philosophical Transactions*, the standard edition put out by the Royal Society itself (hereafter cited: *Phil. Trans.*) and the Hutton, Shaw and Pearson abridgment (hereafter cited: *Phil. Trans. Abgd.*)

The paper of 1716 is by far the most significant for it was wholly concerned with the problem of solar parallax, and it presented a definite plan by which to obtain an accurate knowledge of its value. Moreover, great importance must be attached to the fact that it was Edmund Halley who published the paper. His reputation at the time was firmly established throughout the scientific world; in three years he was to become Astronomer Royal, and his authority on matters astronomical was commanding. His paper was assured of being warmly received: its suggestions constituted a practical program of observation with the promise of definitive results, while the prestige of its author helped to make it a clarion call to scientists everywhere to prepare for the rare opportunity presented by the forthcoming transits of 1761 and 1769. If it is at all necessary, therefore, to select any single factor as the major stimulant in bringing about the enormous enterprise associated with the eighteenth-century transits of Venus, it was certainly Halley's paper of 1716.

IV

A brief discussion of a transit of Venus and its applicability to the problem of evaluating the solar distance seems appropriate at this point. In technical astronomy, discussions of the solar distance make use of the much more suitable term, solar parallax. The word parallax in general means the difference between the direction of a heavenly body as seen by an observer and as seen from some standard reference point. If the earth's center be taken as the standard reference point and a point on the earth's circumference as the observer's position, then the parallax of an object may also be said to be the angular separation of the earth's center and its circumference as seen from the object. In other words, the solar parallax is the angle formed at the sun by lines drawn from the center of the earth and from the observer's station on the earth's surface. It is most useful to define a special kind of parallax when the sun is at the horizon; and this horizontal parallax is simply the radius of the earth as seen from the sun.[54] When we say that the sun's horizontal

[54] Because of the ellipsoidal shape of the earth, the standard solar parallax has been more sharply defined to mean the equatorial horizontal parallax, making use of the earth's equatorial or largest radius as seen from the sun.

parallax is 10 seconds, it is equivalent to saying that, seen from the sun, the earth has an apparent diameter of 20 seconds.

The connection between solar parallax and solar distance can now be established. If for the sake of discussion we place a theoretical observer on the sun, and arm him with an instrument capable of measuring the angular separation between two points and a knowledge of the earth's diameter as well, the matter can be readily clarified. By measuring the angle formed by the earth's diameter as seen from the sun, and knowing its actual length, the problem of determining the distance between the sun and the earth becomes the simple trigonometric one of calculating the altitude of an isosceles triangle where the base length and the angle between the equal sides are known. Conversely, for a terrestrial observer to whom the earth's diameter is known, the problem becomes one of evaluating the angle of parallax.

In practice, the measurement of solar parallax is beset with numerous difficulties and requires, at least from our present point of view, the intercession of one of the inferior planets. The manner in which the planets are used can be seen from an examination of Diagram 1. E represents the earth and its orbit; V Venus and its orbit; S the sun; and AB the line of intersection of the planes formed by the orbits of Venus and the earth. It will be obvious from the inclination of the orbits to one another (actually an angle of about 3 degrees, 23½ minutes) that an inferior conjunction of the two planets will be a rare phenomenon indeed, for it can occur only when the earth and Venus are both present on the same side of the sun along the line of nodes AB. Finally, before leaving Diagram 1, we should note that the earth is always at position A in June and at position B in December, so that all observed transits will occur in those months.

To demonstrate the manner in which a transit of Venus is used to determine the solar distance, let us examine the direct theoretical method which forms the basis for all transit calculations. In Diagram 2, A and B represent two observers on the earth; V the planet Venus in its orbit; aa_1 and bb_1, the paths of Venus across the face of the sun as seen from A and B respectively. Kepler's Third Law, relating planetary periods to their mean distances from the sun, then allows us to determine

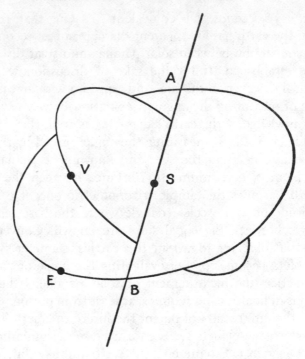

1. THE ORBITS OF VENUS AND THE SUN

the relative position of Venus and the earth with respect to the sun.

Bode's Law, a convenient summary of the relative disposition of the planets in the solar system, provides a simple device for the elucidation of the transit problem. By applying the results of Bode's Law (as shown in the table in Appendix II) to the relative position of the earth and Venus in Diagram 2, we can readily see that the distance of V from the sun is to the distance of E from the sun as 7 is to 10. From this it follows that AV is to Vp as 3 is to 7. Also, on the basis of similar triangles, the spread of the lines of sight at AB will be proportional to the spread at pq. Therefore, if we know the linear distance between A and B, and this is calculable, then we also know the distance pq. For example, if for convenience we assume AB to be 3,000 miles, then pq is 7,000 miles.

At the mid-point of the transit, the centers of Venus as seen

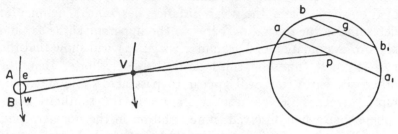

2. VENUS ON THE SUN

on the surface of the sun by the two observers at A and B are represented by p and g. Theoretically our observers at A and B have been able to determine these points as well as the length of the chords aa_1 and bb_1. Consequently they know what proportion the distance pq bears to the diameter of the sun. Suppose they find it to be the 120th part of this diameter. Then because pq is known to be 7,000 miles, the sun's diameter is 120 x 7,000 or 840,000 miles. Now once the actual size of the sun is known, it is not too difficult to calculate its distance from the earth, for only simple trigonometric considerations are involved.

All of this represents an idealized case and, practically speaking, the direct method just described was never actually used. Either too many cumulative errors are introduced in the separate stages of observation or else the observations themselves are impossible to make. The different methods which have been devised to make use of the transits, however, are all variations of this basic approach, largely designed to eliminate error or increase the possibilities of observational success.

V

The method of obtaining the solar parallax which Edmund Halley presented to the astronomical world in 1716 was basic to the transit of Venus observations of 1761 and 1769. Though geographical errors were discovered in the Halleyan proposals before they were put into practice, and some modifications were made in the method itself, Halley's plan remained fundamental to the eighteenth-century calculation of the solar distance.

19

The distinctive feature of Halley's method of using the transit of Venus to obtain the solar distance lies in the manner of determining the separation between the apparent paths aa_1 and bb_1. An examination of Diagram 2 (page 19) will show that the path farthest from the sun's center, bb_1, is shorter than the other, so that Venus will appear to pass across the sun more rapidly when observed from a station in the southern hemisphere than when observed from a station in the northern one. Halley therefore proposed that the time which Venus would appear to spend on the surface of the sun be measured, the instruments required being only a telescope and a good timepiece. The difference in time noted by observers at different stations in the two hemispheres would then supply a means for calculating the difference between the parallaxes of Venus and the sun.

Halley's plan therefore required that both the beginning and the end of a transit be very exactly determined. This in turn demanded four very precise measurements in pairs or phases, the moments of external and internal contact at ingress and egress. As a result of his experience at St. Helena, Halley was of the opinion that the internal measurements at ingress and egress could be easily obtained, for he writes: "I happened to observe with the utmost care, Mercury passing over the sun's disk: and contrary to expectation, I very accurately obtained, with a good 24-foot telescope, the very moment in which Mercury, entering the sun's limb, seemed to touch it internally . . . forming an angle of internal contact. Hence I discovered the precise quantity of time . . . of Mercury . . . within the sun's disk, and that without an error of one single second of time. . . . On observing this I immediately concluded, that the sun's parallax might be . . . determined by such observations. . . . Therefore again and again, I recommend it to the curious strenuously to apply themselves to this observation."[55]

The rotation of the earth naturally introduces additional factors into the problem. Without going into detail, it is merely necessary to indicate that it bears upon the transit problem in

[55] E. Halley, "A New Method of Determining the Parallax of the Sun, or his Distance from the Earth," *Phil. Trans. Abgd.*, VI (1713-1723), 244-246.

two ways. In the first case, it introduces an increment or decrement in transit duration, depending upon the position of the observer and the direction of the earth's rotation. In the second, the rotation of the earth raises the obvious problem of how to place observers in the best possible geographical position to witness the phenomenon. This latter problem assumes special importance in Halley's method. To be effective, his technique required that the duration of the entire transit be measured; and, to make this possible, the widest north-south separation of observers within the cone of visibility of the phenomenon was clearly necessary. The main consideration was to make the difference between observed durations as great as possible in order to keep the errors as small as possible.

In the seventeenth century, the transit of Venus had been observed in pathetic isolation by Jeremiah Horrox and his friend Crabtree. The usefulness of the phenomenon for determining the solar distance was then unknown, and the Enlightenment had yet to intensify and to spread an interest in science. Moreover, in a world which still awaited the unity and certainty of Newtonian mechanics, the question of settling the precise dimensions of the solar system carried with it no sense of urgency. By 1761 and 1769, when the transits were to occur again, the contrast was enormous. The bare mechanism behind the movement of things lay satisfactorily revealed, and the urge to complete the system of the world by discovering its actual dimensions took on a new significance. One measure of that importance is strikingly revealed by the contrast between the limited success of Horrox and Crabtree in merely seeing the transit of Venus, and the catalogue of 120 observers from 62 separate stations who observed it in 1761, and 138 observers from 63 stations who observed it in 1769. In only 15 cases were the same stations used for both events, so that in sum the eighteenth-century transits involved observations from 110 places on the globe.[56] Finally, some of the excitement with which the scien-

[56] S. Newcomb, "Discussion of Observations of the Transits of Venus in 1761 and 1769," *United States Nautical Almanac, Astronomical Papers*, Vol. II, Part 5, pp. 343-345, 358-364, 364-371. These figures do not of course include the large number of unreported observations by interested amateurs throughout the world. Post-transit literature is periodically enriched by the narrative of such observa-

tific world welcomed these rare, astronomical events can be obtained from the not atypical text of C. F. Cassini:

> Elle seule [la circonstance du passage de Vénus] pouvoit dissiper absolument nos incertitudes sur la quantité de la parallaxe du Soleil; elle seule pouvoit fixer avec la derniere précision un élément qui avoit varié jusqu'ici, selon les opinions de divers Astronomes, & selon les différentes méthodes qu'ils avoient employées à sa recherche. Heureux notre siecle, à qui étoit réservée la gloire d'être le témoin d'un événement qui le rendra à jamais mémorable dans les annales des Sciences![57]

---

tions, but by and large they contributed little to the scientific results of the eighteenth-century transits. If they have any importance at all, it is in the limited measure they offer of active scientific commitments by amateurs.

[57] Cassini, *op.cit.*, p. 114.

FRENCH PREPARATIONS FOR THE
TRANSIT OF 1761

THE range and intensity of activity directly connected with the eighteenth-century transits of Venus were, by contemporary standards, enormous. It is quite likely that no other particular scientific problem in the eighteenth century brought so many interests to a single focus as the concern for the solar distance. The mere number of participants directly involved is impressive if bare evidence of that contention.[1] That France should play a leading part in this great international venture is not surprising. Indeed, some aspects of the general cultural and scientific hegemony which France enjoyed in eighteenth-century Europe are clearly discernible from her role in the transit observations.

Outstanding in the task of giving France her position of importance in transit affairs is the name of Joseph-Nicolas Delisle. He deserves special emphasis here for his significant historical connection with Edmund Halley and for his early zeal in the transit enterprise. These factors led him to become a kind of clearing house for correspondence on the transits of Venus.[2]

---

[1] *Supra*, p. 21.

[2] The papers of J.-N. Delisle, or De l'Isle were, in the winter of 1953 when I consulted them, scattered among various depositories in Paris. In the spring of 1953 I learned that the process of assembling them in one place had begun, but I am unaware as to where that unique deposit is to be made.

There is no way of knowing, at this writing, how many of Delisle's literary remains are extant, since his papers are filed together with those of his brothers and his father, the whole constituting an enormous collection of at least twenty folio volumes. These are to be found divided up among the Archives de la Dépôt des Cartes et Plans de la Marine, the Bibliothèque de la Service Historique de la Marine, the Service Hydrographique de la Marine, the Bibliothèque and Archives de l'Observatoire de Paris, the Bibliothèque de la Chambre des Députés, the Archives de l'Académie des Sciences, and the Bibliothèque Nationale. Odd items are also to be found in the archives of the Royal Society of London and in the manuscript collections of the British Museum. No doubt the archives of

At the same time, he made substantial contributions to the original observational technique which Halley had proposed. Much of the French effort to prepare for the transit observations of 1761 can therefore be understood through the work of Delisle.

Considering the incomplete and rather casual treatment which historians of astronomy have given Delisle,[3] a brief sketch of his life and the nature of his scientific activities will not, perhaps, be out of place. This is especially true since his wide-ranging correspondence and extensive contact with fellow scientists throughout the civilized world—a factor which contributed so much to the spread of information about and interest in the transits of Venus—was partly a result of the far-reaching travels he undertook early in life.

The third of four sons, Joseph-Nicolas Delisle was born into a family of historians, geographers, and astronomers in 1688. His formal education was begun by his father and completed at the Collège Mazarin, but not before a penchant for astronomy was revealed by his interest in the total eclipse of the sun for 1706. He was then eighteen years old, and soon

---

the Russian Academy of Sciences also contain much Delisle material, for he spent twenty years there. For additional bibliographical information see G. Bigourdan, *Inventaire général et sommaire des manuscrits de la bibliothèque de l'Observatoire de Paris* (Paris: no publisher, no date), and Charles de la Roncière, "Bibliothèques de la Marine," *Catalogue général des manuscrits des bibliothèques publiques de France* (Paris: Plon-Nourrit, 1907).

[3] Delisle died in 1768, and the *Histoire de l'Académie royale des sciences* (hereafter cited: *Histoire*) for the same year carried his *Éloge* (pp. 320-351) written by Grandjean de Fouchy. Lalande wrote an *Éloge* on his master which was printed in the *Nécrologe des hommes célèbres de France* (Paris: G. Desprez, 1770), Tome v, pp. 1-86. But little interest in Delisle's career was expressed after this until the late nineteenth century, when Petr Pekarski noted his activities in his *Histoire de l'Académie impériale des sciences de Petersbourg* (Paris: n.p., 1870), i, 124-155, and G. Bigourdan drew attention to his manuscripts on deposit in the Bibliothèque de l'Observatoire de Paris. Since then E. Doublet has published *Correspondance échangée de 1720 à 1739 entre l'astronome J.-N. Delisle et M. de Navarre* (Bourdeaux: G. Gounouilhou, 1910); A. Isnard his "Joseph-Nicholas Delisle, sa biographie et sa collection de cartes géographiques à la Bibliothèque nationale," *Comité des travaux historiques et scientifiques. Bulletin de la section de géographie*, xxx (1915), 34-164; and J. Marchand has added "Le Départ en mission de l'astronome J.-N. Delisle pour la Russie (1721-1726)," *Révue d'histoire diplomatique*, 43,1 (October-December, 1929), 1-26. The treatment of Delisle in the *Biographie Universelle* (Michaud) is inadequate, especially in connection with the transit of Venus.

went on to solve several problems in astronomy "par le force de son esprit," that is, before he had any real notion of the subject or any significant training in mathematics.[4] His older brother Guillaume (1675-1726), who was to become first geographer to the king, saw to it that Delisle obtained lessons in geometry, fortification, and mechanics.[5] It was Joseph-Nicolas' first intention to become a surveyor, and with this in mind he perfected his drawing to the point where he soon rivaled his brother in producing maps of precision and elegance. He was never to lose this technique and eventually it was to stand him in good stead in the production of global projections designed to delineate the zones of visibility for transits of Mercury and Venus.

Failing to obtain the first post he applied for, that of Surveyor Royal of Martinique, Delisle turned to the more intensive study of astronomical calculation under the direction of Jacques Lieutaud of the Académie des sciences. By his twentieth year we learn that he has been frequenting the Observatory and that Jacques Cassini has become interested enough in his future to share his time and his labors with him, a significant conquest for a young man only recently turned astronomer.[6] Few important astronomers of the eighteenth century were without their private observatories, and Delisle soon sought and obtained permission to use the cupola of the Palais du Luxembourg. This gave him an observatory but no equipment. The problem of acquiring instruments, which a young, relatively unknown astronomer faced in the early eighteenth century, was of major proportions.[7] Unable to finance the purchase of proper instruments himself, Delisle first tried to meet the deficiency by making wooden instruments, some of which were not too far removed from the earlier models of Tycho Brahe. But these could

---

[4] "Delisle, Joseph-Nicholas," *Biographie Universelle*, x, 334.

[5] Albert Isnard, *op.cit.*, p. 36.

[6] *ibid.*, pp. 36-37. Cassini gave him his manuscript tables of the sun, moon, and planets and allowed him to copy them. The period 1709-1710 also marked the beginning of a long correspondence between Delisle and Cassini, and between Delisle and Cassini's nephew, Maraldi.

[7] In this connection, see the rather fulsome request for such aid which he wrote to the Abbé Bignon. Bibliothèque Nationale (hereafter cited: Bib. Nat.), MSS Fr., 22227, fols. 214-217.

not give him any worthwhile accuracy, and he momentarily gave up the attempt at actual observation to turn, at the request of Cassini,[8] to the production of various astronomical tables.

When he was finally able to equip his observatory with adequate instruments, he quickly fulfilled his early promise and was admitted to the Académie des sciences in 1714 as *élève astronome*. Two years later Delisle was advanced to the rank of *adjoint surnuméraire*.[9] But the pattern of domestic politics among the lords and ladies of the French court soon deprived him of his observatory in the dome of the Palais du Luxembourg. Louis XIV's death when his great-grandson was only five brought the Regency into being under the leadership, for the first eight years, of the Duc d'Orleans (1715-1723). The installation of his older sister, la Duchesse de Berry, at the Palais du Luxembourg, forced Delisle to give up his observatory temporarily. Until her death, he established himself at the private observatory of the Chevalier de Louville at the Hôtel de Taranne.[10] To the precarious livelihood furnished him by giving lessons in mathematics, he soon added a pension of £600 provided out of the fund for foreign affairs. This grant was made to him in recognition of the rather interesting service which he performed for the Count de Boulainvilliers and others in the entourage of the Regent: he supplied them with astronomical calculations applicable to judicial astrology.[11]

An appointment to the chair of mathematics at the Collège Royal in 1718[12] freed him from financial anxieties, and he was able to continue his research in astronomy and to publish regularly. The fame which this important appointment brought him as a teacher and scientist spread rapidly beyond the borders of France, bringing him a wide correspondence with fellow scien-

[8] Isnard, *op.cit.*, p. 37.

[9] *Histoire* (1716), p. 3. Cf. also *Les membres et les correspondants de L'Académie royale des sciences* (1666-1793) (Paris: Au Palais de l'Institut, 1931), p. 69.

[10] C. J. H. Hayes, M. W. Baldwin and C. W. Cole, *History of Europe* (New York: The Macmillan Co., 1949), p. 633, and Isnard, *op.cit.*, p. 37

[11] "Delisle, Joseph-Nicholas," *Biographie Universelle*, x, 335 and Isnard, *op.cit.*, pp. 37-38.

[12] A. Lefranc, *Histoire du Collège de France, depuis ses origines jusqu'à la fin du premier empire* (Paris: Hachette & Cie., 1893), p. 253.

tists everywhere. It also brought him an interesting offer from Peter the Great, repeated several times from 1721 onward, to come to Russia to found a school of astronomy and direct an observatory comparable to the one at Paris.[13] The considerations which Delisle gave to the tempting offer were too prolonged, and Peter died before he could have Delisle's decision, and without having established an observatory in his new capital.

Interest in the transits of Mercury and Venus continued to broaden in the early years of the eighteenth century, stemming not only from the generalized urge to complete the tables of the motions of the planets but also from the increased importance lent to one of them by the publications of Edmund Halley. Between 1723 and 1753 several transits of Mercury took place. Their similarity to those of Venus and the possibility that they might have a comparable use in the calculation of the solar parallax brought them the heightened attention of the scientific world.

Although the incident which had precipitated Halley's essays on the special importance of the transit of Venus had been a transit of Mercury, he saw no reason to stress the utility of the latter in determining the solar parallax. In his opinion the parallax of Mercury itself was too small to yield an accurate value with the available technique. His interest was therefore almost exclusively focused upon the transit of Venus, and he seems to have overlooked even the rehearsal value of the transit of Mercury observations in preparation for the events of 1761. However, this exclusive reliance upon Venus in transit was not, at least in theory, fully echoed by his followers. The advent of the Mercury transit of 1723 moved those who had been excited by the great Halley papers to consider the possibilities of Mercury for determining the solar parallax. At the same time con-

[13] Bibliothèque de la Chambre des Députés (hereafter cited: Chamb. Dép.), MSS 1507, I. This was the period of Peter's attempt to westernize Russia, in the sciences as in all other spheres of action. But in spite of the fact that Peter was a correspondent of the Académie des sciences from his Paris visit of 1717, his interest in pure science should not be exaggerated. The problems of geography and navigation were enormous in the vast reaches of the Russian empire, and Peter saw the practical advantages to be gained from the presence of a school of astronomy and a first class observatory at St. Petersburg. Cf. Isnard, op.cit., p. 38.

sideration was given to the Venus transit of 1761. And by 1723 Delisle was no stranger to that anticipation.

The collection of papers which Delisle brought together on the subject of the transits and his world-wide correspondence— a combination which makes it easy to designate him the un-official archivist of the entire problem—may be said to have begun with consideration of the Mercury transit of 1723. At least this is true insofar as the mute documents are concerned. The earliest of these in his collection at the Observatoire de Paris is a curious broadside[14] by William Whiston (Figure 5). Primarily produced for this 1723 transit of Mercury, and issued only a week or so before the event itself, it has a Janus-like importance. On the one hand, looking backward, it suggested that the exclusive Halleyan concern with Venus unnecessarily limited the usefulness of the transits. Indeed, Whiston felt that the transit of Mercury was superior to that of Venus as a means of determining the solar parallax. In his own words, it was "capable of abundantly satisfying the curious as to that discovery"[15] for a variety of reasons: because its orbital characteristics were better known than those of Venus; because it had already been clearly observed in transit across the solar disk; and because its transit path in 1753 would be closer to the center of the sun than that of Venus in 1761.[16] The key to Whiston's confidence in the superiority of the Mercury transit is perhaps to be found in his acceptance of Halley's own descriptive tables of the planetary motions as wholly sound. The open-mindedness which Delisle maintained on the subject of Mercury and the solar parallax is partially to be explained in the same way. It was not until some time between 1743 and 1747, in a general un-published study on the method of determining the solar parallax, that Delisle began to minimize the usefulness of the Mercury transit, though he still held it possible to calculate the

[14] Archives de l'Observatoire de Paris (hereafter cited Obs. de Paris), Delisle MSS, A.6,8 (60,8,B). The full reference is: William Whiston, *The Transits of Venus and Mercury over the Sun at their Ascending and Descending Nodes for Two Centuries and a Half* (London: I. Senex, 1723).

[15] *ibid.*, fol. 1.

[16] *ibid.*, cf. J.-N. Delisle, *Sur la méthode de determiner la parallaxe du Soleil ou sa distance reele à la terre par l'observation du passage de Mercure ou de Vénus au devant du Soleil*, Obs. de Paris, Delisle MSS, A.6,8 (60,3,B), 9-10.

solar parallax from it.[17] Needless to say, Delisle observed the Mercury transit of 1723, but it did not enable him to make a calculation of the solar parallax. Save for the experience of having observed the transit and the collection of some data on Mercury's movements, Delisle gained little from this event.[18]

On the other hand, looking forward, the Whiston broadside is the first of a long series of printed notices, broadsides, and pamphlets designed to awaken public interest in the importance of the transits. In popular language, these publications explained the role of the transits in determining the distance of the sun from the earth. Supplementing the more standard printed media for such discussion in the eighteenth century, such as the memoirs and transactions of scientific societies, the essays and notices in journals like the *Journal des Sçavans*, the *Journal Encyclopédique*, and *La Feuille nécessaire*, to name but a few, these specialized *fascicules* in popular science publicized the transit problem everywhere. The debate over differences connected with various aspects of the observational problem occasionally reached print in the pages of these journals, further attracting public attention and sharpening public interest in the transits. One such controversy, to be discussed below, between Delisle and Trébuchet, will serve to illustrate the character of some of these debates. Indeed, the minor scandal associated with this Delisle-Trébuchet incident highlights its role in publicizing the transit problem. But behind these public discussions, analyses of the transits and preparations for their observation were under way in the private correspondence of men like Delisle.

Yet correspondence alone was not the only means by which Delisle's growing interest in the transit problem can be meas-

[17] Delisle, *Sur la méthode de determiner la parallaxe* . . . , Obs. de Paris, Delisle MSS, A.6,8 (60,3,B). It is difficult to know when Delisle wrote this essay reviewing the problem, since the 26-page manuscript is undated. But internal evidence, in particular the discussion of the Mercury transits of 1743 and 1753, help to narrow the time of composition down to an interval between 1743 and 1747, when he returned to France.

[18] "Sur le dernier passage attendu de Mercure dans le Soleil et sur celui du mois de Novembre de la présente année 1723," *Mémoires de l'Académie Royale des Sciences* (hereafter cited: *Mémoires*), 1723, pp. 105-110, and "Observation du passage de Mercure sur le Soleil; faite à Paris dans l'Observatoire royal, le 9 Novembre 1723, au soir," *Mémoires* (1723), pp. 306-343.

ured. In 1724, with the Mercury transit of 1723 behind him, Delisle had the opportunity of accompanying a friend on a trip to London. There he was warmly welcomed by Halley, whose confidence in Delisle was displayed by the fact that he gave the French astronomer a copy of his unpublished astronomical tables on the sun, moon, and planets, tables which became available to the public at large only in 1749, seven years after Halley's death.[19] Halley was evidently much impressed with Delisle; for long after the visit, others wrote Delisle to say that Halley "very frequently speaks of you with great respect."[20] Delisle was no less amiably welcomed by Newton, who presented him with a copy of his portrait as a souvenir of their meeting.[21]

It seems that Delisle and Halley, if not Newton as well, very likely discussed the transit problem in terms of the location of stations of observation and the duration of the phenomenon upon the basis of Halley's tables. The dissatisfaction which Halley felt with regard to the tables was apparently discussed at the same time, for years later Delisle referred to the fact in a letter to an English correspondent.[22] This was one reason why the tables were not published until some time after Halley's death, for it had been his intention to perfect them before publication.[23] The importance of this meeting in 1724 for the transits of Venus in 1761 and 1769 is therefore very great, for Delisle later published corrections of Halley's predictions— an act which led to acrimonious debate over the discovery of the original errors and their corrections.

The year following his voyage to London became decisive in Delisle's career. The death of Peter the Great did not bring

[19] Isnard, *op.cit.*, 39, and "Delisle, Joseph-Nicholas," *Biographie Universelle*, x, 334. This trip also resulted in an invitation to membership in the Royal Society.

[20] Archives du Dépôt des Cartes et Plans de la Marine (hereafter cited: Dép. Marine), Delisle MSS, Vol. 115, XVI-6, fol. 3.

[21] The Newton portrait has disappeared from the Delisle papers, so that its character and quality, whether oil copy or engraving, cannot be determined.

[22] British Museum, MSS 531, e. 47, fol. 14.

[23] "Lettre de M. de Lisle, Professeur Royal . . . de l'Académie royale des sciences a M. . . . sur les tables astronomiques de M. Halley," *Mémoires de Trévoux* (Février 1750), p. 377. This same letter was inserted in the *Journal des Sçavans* for December 1749.

to an end Russian interest in Delisle. In the brief, loose-moralled and ugly-mannered career of Catherine I as Tsarina (1725-1727), a mote of grace may be granted her for the renewal of Peter's invitation to Delisle to come to Russia. This time he accepted, but in order to guard against the loss of his post, he sought and obtained a royal warrant or patent which authorized his departure and guaranteed his position as academician and professor for an absence not to exceed four years.[24] Authority was further granted him to choose assistants for his task, and he chose his brother, Louis Delisle de la Croyere, astronomer and member of the Académie,[25] and an instrument maker named Vignon,[26] whom Delisle obtained from the atelier of Louis Chapotot.[27]

The details of Delisle's stay in Russia remain part of the untold story of the Eastward movement of modern science, one facet in the mosaic of Russian Westernization. Delisle's four years in Russia turned into twenty-two, and to that country he gave "ses lumières, ses réflexions et ses ordres."[28] In return, he was well rewarded financially, and his every material need supplied. The trip to St. Petersburg itself was a long leisurely excursion, a kind of grand tour through the centers of European astronomy, to visit former correspondents and make new acquaintances, and to acquire instruments and books for the new observatory. A stop in Danzig, for example, enabled him to acquire the correspondence of Hevelius and the journals of his observations from 1657 to 1686.[29] His mission to train the Russian empire's astronomers met with success, and in very little time the school of astronomy at St. Petersburg acquired a fine reputation. Delisle seems to have taken great pains in training his students, writing elementary treatises for teaching purposes,

[24] Bib. Nat., mss Fr., 9678, f 31, fol. 17.
[25] ibid., fol. 18, and Chamb. Dép., mss 1508, 1.
[26] Bib. Nat., mss Fr., 9678, f 31, fol. 18.
[27] Maurice Daumas, Les instruments scientifiques aux XVIIe et XVIIIe siècles (Paris: Presses Universitaires de France, 1953), pp. 103-105. Chapotot's was one of the more important French ateliers for scientific instruments of the late seventeenth and early eighteenth centuries.
[28] Bib. Nat., mss Fr., 9678, f 31, fol. 30.
[29] Chamb. Dép., mss 1507, 1, 36. Letter from Delisle to Blumentrost, sometime physician to Peter the Great and president of the St. Petersburg Academy of Sciences.

furnishing them with books and instruments, and helping them to get financial grants.

Frequent travel throughout Russia enabled him to bring together a large amount of useful physical and geographical information, most of which he brought back to France when he returned in 1747.[30]

Partly in recognition of his services and partly because of the inherent value of his geographical and astronomical collection, his library was purchased by the French government a few years after his return. In return, he received the title of *Astronome de la Marine* and a life annuity of £3,000 for himself, £600 for his secretary, and £500 for his pupil and assistant,[31] Charles Messier.[32] Until 1768 when he died, Delisle continued to observe from the Hôtel de Cluny, using as his observatory[33] the solid, octagonal tower which today forms the entrance to the museum.

With his interest in the transits reinforced by the meeting with Halley in 1724, Delisle undertook a thorough study of the problem of solar parallax. His analysis took two distinct directions. First he approached the problem historically by examining various past attempts to determine the solar distance, probably with the intention of publishing a treatise on the subject. His research into this aspect of the problem, while incomplete in detail, reveals the extent of his ambition, for he had gone back to the pre-Socratics and forward as far as his contemporaries, discussing the means employed, the results obtained, and the degree of satisfaction which they gave.[34] This work, lying undated in the archives, was very likely completed

---

[30] "Delisle, Joseph-Nicholas," *Biographie Universelle*, x, 335.

[31] Doublet, *op.cit.*, pp. 7-8. The library itself was originally deposited, about 1750, in the Dépôt de la Marine, where it remained until 1795. By a decision of the Comité de Salut Public it was, for some obscure reason, divided between the Dépôt de la Marine and the Observatoire, a division which continues to plague contemporary research. Cf. C. de la Roncière, *op.cit.*, pp. vi-vii.

[32] On Messier and Delisle as well as Messier's later discoveries, see O. Gingerich, "Messier and His Catalogue," *Sky and Telescope*, xii, 10 (August 1953), 255-258, 265.

[33] *ibid.*, p. 256. Messier used the same observatory until 1817, and for a time, Lalande also lived there.

[34] His research notes are collected under the title *Histoire des opinions des astronomes sur la distance du Soleil et de la Lune à la Terre; et des differens moyens qu'ils ont employés a connoître ces distances, & ce.*, Obs. de Paris, A.6,9 (61,20,A-T).

by the middle of the century, not long after his interest in the history of science was revealed by the publication of an anthology of brief memoirs on the history and progress of astronomy, geography, and physics.[35]

The second manner in which Delisle approached the problem was to analyze, in the unpublished set of notes referred to earlier,[36] the data compiled and the methods suggested by his scientific contemporaries. In addition to Halley and Whiston, Delisle discussed the contributions of Cassini and Maraldi, hastening to point out that they were unsuccessful in deducing an adequate figure for the solar parallax from a Mercury transit.[37] Russian research, under his direction, was equally unsuccessful, and it was this "ce qui . . . [l'] a fait ecrire en Portugal, en Italie en France et dans le nord," in order to discuss his difficulties with fellow scientists and at the same time share his new ideas on transit observations.[38]

The technique which Delisle proposed was really a simplification of Halley's method.[39] With Halley, Delisle assumed that

[35] *Mémoires pour servir à l'histoire et au progrès de l'astronomie, de la géographie et de la physique . . . de l'Académie royale des sciences de Paris et celle de Saint Petersbourg, qui n'ont point encore été imprimées, comme aussi de plusieurs nouvelles, observations et réflexions rassemblées pendant plus de 25 années* (St. Petersbourg, 1738). Delisle's interest in the history of astronomy actually goes back to a much earlier date. In 1728 he began to prepare an *Abrégé de l'histoire des observations astronomiques* to be read at a public assembly of the St. Petersburg Academy, but he never completed it, carrying the story only as far as Albugensis. Nevertheless, 36 years later, in 1764, he made additional entries on Chinese astronomy to these notes for the *Abrégé*. And, it should be noted, there is also much material to be found here on Arabic astronomy. Obs. de Paris, A.7,10, No. 8.

[36] See note 16.

[37] Obs. de Paris, Delisle MSS, A.6,8 (60,3,B), fol. 12.

[38] *ibid.*, fol. 14.

[39] "Sur le dernier passage attendu de Mercure dans le Soleil, et sur celui du mois de Novembre de la présente année 1723," *Mémoires*, 1723, pp. 105-110; "Sur la methode de determiner la parallaxe du Soleil ou sa distance reele à la Terre par l'observation de Mercure ou de Vénus au devant du Soleil," Obs. de Paris, Delisle MSS, A.6,8 (60,3,B), fols. 16-21; and in "Extrait d'une lettre de M. Delisle, écrite de Petersbourg le 24 Août 1743, & adressée à M. Cassini, servant de supplément au Mémoire de M. Delisle, inséré dans le volume de 1723, p. 105, pour trouver la parallaxe du Soleil par le passage de Mercure dans le disque de cet Astre," *Mémoires*, 1743, pp. 419-428. Particularly significant, and further proof of his early interest in the problem is the statement made in the above manuscript (fol. 10) concerning his method: "Plus d'un an avant le passage de 1723 j'avois inventé une nouvelle methode pour calculer les passages de mercure sur le Soleil. . . ."

that exact moment of contact could be determined, but he recognized that observations of the entire transit, or at least of ingress and egress, were not necessary; that in fact, this required excellent weather conditions for the whole duration of the transit—a factor which could not entirely be relied upon. Consequently it occurred to him that the observer who sees the planet take the longer path across the sun witnesses the beginning of the transit much earlier than the observer who records the shorter path. This is rather obvious: since both are apparent paths, the longer route must have an earlier beginning and a later ending than the shorter one. It was, therefore, necessary only to record the local time for the exact moment of contact at either ingress or egress. Then, from an exact knowledge of the longitude at the place of observation, the local time could be adjusted to Greenwich or Paris time. The difference in time between two points of observation would then be proportional to the time by which the contact for the longer path precedes the contact for the shorter path, and hence proportional to the distance between them. Once this is known, it follows from the basic method described in Chapter 1 that the size and distance of the sun can be obtained.

From the point of view of the observational problem, Delisle's method was obviously superior to Halley's. It required only one observation from a given point and thereby reduced the likelihood of failure due to adverse weather conditions; or if observations of the planet's ingress and egress were both made at the same station, the probability of success was doubled. The range of useful stations was also extended by this method to include those parts of the world where only the beginning or the end of a transit would be visible. But Delisle's scheme was not all milk and honey, for the requirement of an extremely precise determination of longitude at the place of observation was, in spite of all the advances made in that direction during the eighteenth century, a problem of no small proportions. However, it was not methodology alone which concerned Delisle, but also the practical effect of the power and quality of the lenses employed in various telescopic combinations upon the apparent time spent by the planet on the face of the sun. From his correspondents Delisle requested, and received, confirmation of the idea that this apparent duration varies directly with the length of the telescope, and that this effect must be understood before the difference in time due to the geographical location of the observer's station could be properly utilized.[40]

From such correspondents as Dirk Klinkenberg, the "Curieux Geometre et Astronome de Harlem,"[41] Delisle also found support for the idea, earlier expressed by Whiston and others, that the solar parallax, in spite of a succession of failures between 1723 and 1743, could still be deduced from the transits of Mercury.[42] Though it remained virtually unquestioned that the transit of Venus of 1761 offered the greatest opportunity for arriving at a definitive value of the solar distance, Delisle and the astronomical confreres within reach of his pen thought it worthwhile to emphasize the transit of Mercury of 1753 as the best of all Mercury transits[43] before the year 1761, when Venus would at last make her rare appearance upon the solar disk.

Before the transit of Mercury of 1753 took place, however, the French attempted to determine the solar parallax independently of the transit technique. This enterprise, under the leadership of Nicolas Louis de la Caille, produced an expedition to the Cape of Good Hope[44] and a measure of international cooperation that is a preview in miniature of the astronomical activities to be associated with the Venus transits of 1761 and 1769.

Having acquired much experience in the cataloguing of stars visible at the latitude of Paris, La Caille was called upon by the Académie des sciences to perform a similar task by charting

[40] Obs. de Paris, Delisle MSS, A.6,8 (60,3,B), fols. 17-18.
[41] ibid., fol. 17.
[42] Dirk Klinkenberg, Verhandeling Over het vinden van de Parallaxis der Zon; Zynde eene Beschryving hoe de afstand, tusschen de Zon en de Aarde kan gevonden worden Door den schynbaaren weg der Planeeten Venus en Mercurius over de Zon: Nevens de Afbeeldingen van drie zulke Verschynsels, welke voorvallen zullen, het eene in het Jaar 1743, de ander 1753, en de derde in't Jaar 1761 (Haerlem: Jan Bosch, 1743).
[43] Whiston, op.cit., fol. 1, had thought the same thing; and Klinkenberg, op.cit., p. 40, had even thought that the transit of Mercury of 1753 was not less favorable than that of Venus for 1761 for determining the solar parallax. Cf. Delisle, Obs. de Paris, Delisle MSS, A.6,8 (60,3,B), fol. 21.
[44] N. L. La Caille, "Relation abrégé du voyage fait par ordre du Roi, au Cap de Bonne-Espérance," Mémoires, 1751, pp. 519-536.

the skies of the southern hemisphere.[45] The location of most of the brighter stars in the northern skies had been fairly well determined, but those of the south were virtually unknown. Halley, as a result of the St. Helena expedition, was one of the few professional astronomers to have looked at the southern sky. In effect it was La Caille's job to fill in the blank spaces in the planisphere between the northern constellations of Ptolemy's Almagest and the handful described by the early Portuguese navigators of the southern oceans.[46] At the same time, the Académie desired that he make these observations from a place where he might also determine the parallax of the moon and, at the occasion of the opposition of Mars and the conjunction of Venus, make new attempts to determine the solar parallax. With this purpose in mind, and having obtained the agreement of the "States General of Holland"[47] and the permission of the Dutch East India Company, in whose possession the colony then was, he set sail for the Cape of Good Hope in the autumn of 1750.[48]

On the eve of his departure, in a widely circulated notice, the *Avis aux Astronomes*, he wrote that "parce qu'on ne peut parvenir a la determination exacte des parallaxes, que par des observations concertees, & faites en meme tems aux deux extremites d'un arc du meridien, j'invite tous les Astronomes fournis des instrumens convenables, a prendre part a ces recherches, si interessantes pour le progres de l'Astronomie & de la Navigation."[49] This invitation also included detailed instructions as to the actual observations, divided into sections on the lunar parallax, the parallax of Venus, and the use of various appro-

[45] La Caille, *Avis aux Astronomes*, Obs. de Paris, MSS A.6-9 (61,5,A).

[46] D. S. Evans, "La Caille: 10,000 Stars in Two Years," *Discovery*, XII, 10 (October 1951), 318-319. While engaged in this task, La Caille took issue with Halley for inserting the constellation "Robur Carolinum" (King Charles's Oak Tree) in the southern heavens. This sort of flattery had no place in science and he therefore drafted his own constellations. A true son of the Enlightenment, his were named after the sciences and the liberal arts: *Antlia Pneumatica*, the air pump; *Fornax Chimica*, the chemical furnace; *Apparatus Sculptoris*, the sculptor's studio; and so on. D. S. Evans is a member of the staff of the Royal Observatory at Cape Town.

[47] La Caille, *Avis aux Astronomes, op.cit.*, fol. 1.

[48] Evans, *op.cit.*, p. 315.

[49] La Caille, *Avis aux Astronomes, op.cit.*, fol. 1.

priate tables.[50] While not universal, the response to La Caille's invitation was significant: Bradley observed at Greenwich, Grischov in Russia, and Lalande in Berlin, the most important station of all, for Cape Town was almost on the same meridian as Berlin.[51] French scientific ties with Berlin were very close, for Maupertuis was then president of the Berlin academy of sciences, and Lalande was welcomed with alacrity.[52] This was probably the first time in the growth of astronomy that an attempt was made to get widely-separated observers to make simultaneous observations of the same celestial bodies. At least, this was Lalande's impression.[53]

The work which La Caille performed at the Cape in two years was prodigious. It immediately raised him to the first rank among astronomers and brought him permanent fame. Present-day astronomers with an interest in the southern skies consider him the father of southern astronomy.[54] But his other efforts, while of a high order of excellence, were not equal to his achievement as a stellar cartographer. An error inherent in the location of his observatory produced the disquieting result that the earth was a prolate rather than an oblate spheroid,[55] a matter virtually settled some time before, and his calculation of the solar parallax was also unsatisfactory.[56] The dissatisfaction, it

[50] ibid., fols. 2-3.   [51] Evans, op.cit., pp. 317-318.
[52] P. Brunet, Maupertuis (Paris: Libraire Scientifique Albert Blanchard, 1929), I, Étude Biographique, 135-138; II, L'Oeuvre et sa place dans la pensée scientifique et philosophique du XVIIIe siècle, 180-181. The enthusiasm with which Maupertuis met Lalande was more than an expression of scientific comity or narrow patriotism, for he was vitally interested in the problem of lunar parallax as it bore upon the question of the figure of the earth, that first great problem of the post-Newtonian epoch in which Maupertuis had earlier played such an important role and which he was then reconsidering in his Lettre sur le progrès des sciences. ibid., II, 181. As a matter of fact, Maupertuis took advantage of Lalande's presence to get a considerable and varied amount of work done at the Berlin Observatory above and beyond the cooperative observations with La Caille. ibid., I, 137-138. See also the Procès-Verbaux de l'Académie royale des Sciences (hereafter cited: Proc. Verb.), 1753, fol. 14, for the report by Bouger, de Thury, and Le Monnier on this aspect of Lalande's work.
[53] See Lalande, "Observations faites par ordre du Roi, pour la distance de la lune à la Terre, a l'Observatoire royal de Berlin, en 1751 & 1752," Mémoires, 1751, pp. 457-480.
[54] Evans, op.cit., p. 315.   [55] ibid., p. 319.
[56] ibid., p. 318. La Caille's results are to be found in "Mémoire sur la parallaxe du Soleil, qui résulte de la comparaison des observations simultanées de Mars & de Vénus, faites en l'année 1751 en Europe & au Cap de Bonne-Espérance,"

should be pointed out, was not universal among his contemporaries, for the result was more or less consistent with earlier attempts to determine the solar parallax in this manner.[57]

There is, however, a prophetic comment in the concluding remarks which La Caille made in the *Avis aux Astronomes* before his departure for South Africa. It is of the utmost importance for the ultimate evaluation of the solar parallax deduced from the transits of Venus, and it is therefore worth quoting at length. Referring to the last of the Halley memoirs on the transit of Venus, he says:

M. Halley conclud que par le passage de Vénus sur le disque du Soleil en Juin 1761, on pourra déterminer la parallaxe du Soleil à 1/500 près; pourvu qu'on observe ce passage dans . . . certaines circonstances de tems & de lieux. . . . Mais quelque déférence que j'aye d'ailleurs pour les sentimens de ce grand homme, cette précision me paroît absolument impossible. Car quand même il arriveroit par le plus grand hazard du monde, qu'un Astronome bien exercé, placé vers l'extrémité boréale de l'Amérique, eût le bonheur de voir l'entrée de Vénus sur le . . . Soleil près de son coucher, & le lendemain sa sortie hors du disque du Soleil levant, je ne puis croire qu'il lui fût possible d'en déterminer les instans à 2 secondes de tems près, comme M. Halley le suppose. . . . 1. Parce que les bords du Soleil voisin l'horizon sont dans une ondulation continuelle, & que les réfractions irréguliers qu'il souffre, font paroître à tout moment comme de petites portions qui se détachent du disque du Soleil. 2. Parce que le mouvement rapide du Soleil & de Vénus dans le champ d'une lunette qui grossit beaucoup, rend très difficile la détermination exacte du moment du contact de leurs limbes. Mais un fait bien constant mettra la chose hors de doute. Le mouvement horaire apparent de Mercure dans son noeud ascendant & sur le disque du Soleil, est à celui de Vénus, comme 3 à 2, & par conséquent on doit déterminer l'instant du contact intérieure des limbes du Soleil & de Mercure

---

Mémoires, 1760, pp. 73-97. It should be pointed out that though the memoir is to be found in the volume for 1760, it was not actually published until 1766.

[57] See C. F. Cassini, *Histoire abrégé de la parallaxe du Soleil, op.cit.*, 123-124. ". . . [L]a méthode de chercher cette parallaxe par celle de Mars . . . meritoit quelque confiance. . . . [T]outes les fois que cette méthode a été employée par un observateur habile & dans les circonstances favorables, elle a toujours donné des resultats peu differents entre eux; . . . en 1704 . . . Maraldi . . . la [la parallaxe du Soleil] trouvé de 10″. Quinze années apres . . . mêmes operations de la part de M. Maraldi, même resultat encore. Enfin M. de la Caille, . . . apres avoir discuté, pesé, & calculé toutes . . . observations, finit par conclure la parallaxe du Soleil de 10″ 1/5."

avec plus de précision que celui du Soleil & de Vénus. Or en 1743, le Ciel étant fort serein, . . . des Astronomes des plus habiles qui observerent avec d'excellens Télescopes le contact intérieur de Mercure & du Soleil, different beaucoup entre eux; & la différence alla à plus de 40 secondes de tems. . . . [Mais] . . . ce que je dis ici, n'est pas pour m'ériger en censeur des Ecrits de M. Halley, dont j'honore infiniment la mémoire, ni pour diminuer l'idée que l'on a toujours eue de l'utilité de l'observation de ce fameux passage de Vénus: mais seulement afin que l'autorité de [ce] célèbre Astronome ne serve pas de prétexte, pour faire négliger l'occasion de déterminer la parallaxe du Soleil, par les observations que je propose.[58]

La Caille was of course absolutely right in suggesting that no occasion be neglected for determining the solar parallax; and he spoke with justification against the false security of relying upon the authority of one man in science, however great. But we have already seen that the transit of Mercury was not being neglected, in spite of Halley's position on the matter, and that preparations were going forward to make use of it in 1753, when Mercury would be in the most favorable position in its observed history to be of use for the solar parallax. The decade before the Venus transits of 1761 and 1769 was therefore to witness two attempts at an evaluation of the solar parallax: that of La Caille and his associates and the much more direct effort involving the transit of Mercury of 1753. Both made their contributions toward the great enterprise of the Venus transits, not only in the pattern created for bringing about cooperative undertakings, but also in the very valuable scientific experience gained.

In his criticism of the degree of accuracy to be expected from the transits of Venus, La Caille stood splendidly alone. Delisle had earlier mentioned the problem of lenses and telescopes and their effect upon accuracy, but this was merely a precautionary measure, or at best criticism of a constructive order. Full discussion of La Caille's careful criticism of the transits of Venus is not appropriate at the moment, and will be undertaken at another place, but it is worthwhile to point out here that it was substantially sound and more than just. Moreover, his was the first comment to raise public doubts as to the possibility of

[58] La Caille, *Avis aux Astronomes, op. cit.*, fol. 4.

fulfilling the goal for which extraordinary interest in the transits of Venus was proclaimed. Incidentally, Delisle took an effective part in the circulation of La Caille's ideas and the importance of the kind of observations which he advocated. His letters on the subject went to Sweden, England, and Italy,[59] while those sent to his German correspondents were deemed important enough for public circulation.[60] In preparation for the same event he presented to the Académie in September 1750, a diagram showing the apparent path of Mars against a background of stars important to the observations in question (Figure 6).[61] When the observations themselves were completed, Delisle was "contenté de receueillir toutes les observations qu'on fait tous les astronomes de l'Europe, que j'avois excitée à cela en leur envoiant l'avis de M. de la Caille."[62] At the same time he prepared a report[63] on the observations which made its way across Europe through his innumerable letters, frequently in conjunction with an announcement of the forthcoming transit of Mercury and instructions as to how to observe it.[64]

In a very important study on scientific method published in the last quarter of the nineteenth century,[65] while speaking of the problem of obtaining accurate measurements of natural events, W. Stanley Jevons illustrated his point with the example of the transits of Venus. "The sun forms a kind of back-

[59] Discussed in letters to: Eustache Zanotti, astronomer at the Institute of Bologna, Obs. de Paris, B.1,1-8 (140), No. 13; M. Perelli, Professor of Mathematics at Pisa, ibid., No. 15; R. P. Ximenez, Jesuit Professor of Astronomy at Florence, ibid., No. 16; and Birch, Secretary of the Royal Society, ibid., No. 23.

[60] These letters were translated and printed for distribution. See "Zwei astronomische Schreiben an den Professor Bose in Wittenberg," Obs. de Paris, B.1,1-8 (XIV et XV), Nos. 144.72, 144.73, and 144.74. Also "Des Herrn de l'Isle Circularschreiben an alle Astronomen in Europa," ibid., No. 144.75. Letter dated "Paris, 29 janvier 1751."

[61] From a copy in the Obs. de Paris, Delisle MSS, A.6,9 (61,3,B).

[62] Letter to Bevis dated Paris, 30 November, 1752, Obs. de Paris, B.1,1-8 (140), No. 23.

[63] "Recherche de la parallaxe de mars par les observations d'Europe correspondantes a celles qui ont été faites au Cap de Bonne Esperance par M. l'abbé La Caille pendant l'année 1751," Obs. de Paris, A.6,9 (61.1).

[64] There are some 266 letters almost exclusively concerned with this subject in the archives of the Paris Observatory, B.1,1-8 (140). In addition to those already indicated (note 59 above), the correspondence includes names like Wargentin in Stockholm (No. 59), Marinoni in Vienna (No. 57), Koenig at The Hague (No. 76), and Sabatelli in Naples (No. 65).

[65] W. S. Jevons, The Principles of Science: A Treatise on Logic and Scientific Method (3rd edn.; London: Macmillan and Co., 1879).

ground," he wrote, "on which the place of the planet is marked, and serves as a measuring instrument free from all the errors of construction which affect human instruments. . . . It has been sufficiently shown that by rightly choosing the moments of observation, the planetary bodies may often be made to reveal their relative distance, to measure their own position, to record their own movements with a high degree of accuracy."[66] This kind of measurement by natural coincidence is of course the essence of the technique for utilizing the transits of both Mercury and Venus. But the question of "rightly choosing the moments of observation" lies at the heart of the problem. In the eighteenth century, that century so fully devoted to reducing all natural phenomena to principles of order and regularity, the answer to that question lay in the hieroglyphic of astronomical tables and in the carefully arranged data of innumerable observations designed to fix the sweep of a planet in space and time. The repeated re-evaluation of planetary tables is therefore a major preoccupation of eighteenth-century astronomy. Indeed, the astronomical table is at once the springboard for criticism and its improvement the considered result of all action. The attempt, therefore, by the proper selection of time and place, to turn sun and planet into an instrument for determining the real distance between each, must originate in the tables which dictate that choice.

The French approach to the transit of Mercury in 1753, like their approach to that of Venus in 1761, can be said to begin with Delisle's consideration of Halley's tables after their publication in 1749. Delisle commented on the excellence of these tables and their general value but he also made haste to add several necessary corrections.[67] The labor behind this task was of the highest caliber and the corrections themselves were sufficiently important to elicit compliments from even so hostile a later critic as Delambre.[68]

[66] *ibid.*, pp. 294-295.
[67] See "Lettre de M. de Lisle . . . sur les tables astronomiques de M. Halley," *Mémoires de Trévoux* (Février 1750), pp. 377-380, and *Lettre de M. De L'Isle, . . . sur les tables astronomiques de M. Halley* (Paris: Imprimerie de Quillan, 1749).
[68] Delambre, *Histoire de l'astronomie au XVIIIe siècle* (Paris: Bachelier, 1827), pp. 325-326.

The publication of Halley's astronomical tables[69] and the approaching transit of Mercury of 1753 precipitated a spate of letters and bulletins from Delisle to the principal astronomers of Europe and even the new world.[70] Most of this activity seems to have begun in the fall of 1752 and continued up to the transit itself on 6 May 1753. As an immediate forerunner of the transit of Venus of 1761, it is indicative of the pattern of things to come. Once again, the picture can be most clearly seen through the activities of Delisle.

In the autumn of 1752 Delisle wrote to the Governor-General of Canada, the Marquis de la Galissonière, who was then in Paris. It would be advantageous, he wrote, to observe the transit of Mercury from Quebec, the island of St. Domingue, and Cayenne; and with this in mind he asked the Marquis to assist him in getting observations made in these parts. In a letter to La Galissonière written from Rheims, 19 October 1752, Delisle thanked him for undertaking to send his memoir to the Reverend Father Bonnécamp, a Jesuit Professor of Hydrography at Quebec, apologizing for the lateness of its delivery "comptant toujours qu'il seroit assez Temps pur ecrire a Kebec avant le 6 may prochain."[71] He renewed the earlier request for observations at St. Domingue, stressing the additional purpose of determining "d'une manière sûre La Longitude de cette isle." Apparently, it was this very problem which led La Galissonière to support Delisle's suggestion that Cayenne be used, "parceque La Longitude de ce Lieu étant assez bien connüe par les dernieres observations qu'y a fait Mr. de la Condamine on pouvait s'en servir pour determiner la parallax du Soleil, ou confirmer celle que l'on deduira des observations de Mr. de La Caille, en comparant l'observation de Cayenne sur le prochain passage de Mercure . . . avec Celles qui se feront dans Les indes orientales et en europe. Je me suis assuré de celles des indes

[69] They were to be republished with corrections and additions in 1752 by one of the principals in the transit of Venus observations, Chappe d'Auteroche: "Éloge de M. l'Abbé Chappe," *Histoire*, 1769, p. 164. A second translated edition was put out by Chappe again in 1754; Delambre, *op.cit.*, p. 303.
[70] See I. B. Cohen, "Benjamin Franklin and the Transit of Mercury in 1753," *Proceedings of the American Philosophical Society* (hereafter cited: *Proceed. Am. Phil. Soc.*), Vol. 94, No. 3 (June 1950), 222-232.
[71] Obs. de Paris, Delisle Correspondence, B.1,1-8 (140), No. 1.

orientales; ayant ecrit pour cet effet aux Missionaires Jesuites
. . . a pondichery, Chandernagor, Macao et pekin."[72] The problem as Delisle saw it was to find someone at Cayenne or in
the neighboring region who would be capable of making the
observations. But assuming that the interest of so powerful a
figure as the Marquis de la Galissonière would obtain the services of an observer, he included in the same letter a detailed
set of instructions for observing the transit of Mercury at Cayenne. By the ninth of November, when he heard from La Galissonière again, that broad assumption was nullified, for La Galissonière wrote that his instructions (like those for Quebec) had
already been sent, but that it was unlikely that there would be
anyone at Cayenne capable of following them.[73]

The Delisle memoir destined for the Jesuit observer took a
rather interesting path in getting to Quebec. Recognizing that
it was too late to attempt to get the instructions to Quebec by
the usual avenues of communication between France and her
North American colonies, La Galissonière wrote to the Governor of New York with an unusual request. The brief letter
is worth quoting in its virtual entirety.

Paris, October 10, 1752

Sir,

The Astronomers here have a little too late proposed to me, to get
at *Quebeck*, Observations made of the Passage of *Mercury* upon the
Sun, which should happen the Beginning of *May* next; I cannot do
this, which concerns all Nations, who desire the Perfection of Geography and Astronomy, but by a Jesuit Professor of Hydrography, at
Quebeck; but it is impossible that he should in time receive the Letter . . . thro' our Colonies; and I have no hopes of his receiving it,
unless you will forward it by *Boston*, or *New-York*. I send you Dupli-
cates of it, that you may send by different Vessels; and I send it
open to you, that you may see that nothing in it concerns the Government.

The Taste I know you have for everything that may be useful and
agreeable to Mankind, makes me hope that you will favour the
Passage of that Letter.[74]

[72] *ibid.*   [73] *ibid.*, No. 9.
[74] Cohen, *op.cit.*, pp. 230-231. This letter is to be found in a rare pamphlet,
*Letters Relating to the Transit of Mercury over the Sun, which is to happen
May 6, 1753*, printed by Franklin. A facsimile of it is contained in Cohen's article,

The Governor did favor the passage of the letters dispatching them by "express overland to Quebec,"[75] but not before he had seen to it that a translation was made under the supervision of James Alexander of New York. In turn, Alexander sent copies of the translated memorial and letters to Cadwallader Colden and Benjamin Franklin, who recognized their importance and struck off fifty copies for American distribution.

Though the news about the transit of Mercury was not entirely unexpected,[76] the translation and publication of the letters of La Galissonière and Delisle had at least one important consequence. James Alexander immediately saw the significance of the event for the later, and more important, transit of Venus. With that in mind, he wrote Franklin on 29 January 1753 to say:

. . . that that which had baffled all the Art of man hitherto To discover with any Tolerable Certainty (Viz: the Sun's Distance from the Earth) may with great Certainty be Discovered by the transit of Venus over the sun the 26 of May old stile if well Observed in the East Indies and here and there Observations compared Together.

It would be a great honour to our young Colleges in America if they forthwith prepared themselves with a proper apparatus for that Observation and made it. Which I doubt not they would Severally Do if they were Severally put in mind of it and of the great Importance that that Observation would be to Astronomy. . . .[77]

A second letter written to Cadwallader Colden to accompany the translated letters shows even more clearly what Alexander meant by discussing the transit of Venus when the transit of

---

pp. 229-232, together with the interesting story of how it came to be printed by Franklin. I am indebted to the author for calling my attention to the story several years ago.

A parallel set of instructions written for the island of St. Domingue (the present-day Dominican Republic) by Delisle are also available at the Obs. de Paris, B.1,1-8 (140), No. 19.

[75] From a letter by Franklin to James Bowdoin, dated Philadelphia, 28 February 1753. Quoted by Cohen, *op.cit.*, p. 223.

[76] The transit of Mercury for 1753 was announced in *Poor Richard Improved . . . for 1753*. See F. L. Mott and C. E. Jorgenson, *Benjamin Franklin: Representative Selections* ("American Writers Series," [New York: American Book Co., 1936]), pp. 225-260.

[77] Collections of the New York Historical Society (1920), Colden Papers, Vol. 4, pp. 367-368.

Mercury was imminent. Alexander asked Colden to "assist in prepareing things for . . . observing the transit of Mercury . . . and in makeing the observation, for Except your Self & me, I believe there's none in the province any way acquainted with observations of that kind, and observing that transit *might show some young men how to observe the transit of Venus in 1761*"[78] The usefulness of the transit of Mercury as a training device in preparation for 1761 was therefore, in America at least, immediately recognized. It was a prescient conclusion in the light of later events. Furthermore, it was made obvious that activities connected with the transit of Mercury in 1753 would also serve to stimulate interest in those of Venus in the next decade.

II

Private correspondence, as has already been suggested, played a large role in the discussion of scientific problems like those connected with the transits. At times, however, their private character was completely nullified by their appearance in print with or without the knowledge of the correspondents. The Franklin printing just discussed and the German edition of Delisle's letters referred to earlier are examples of this practice. The quality of the letters in detail and continuity is of a very high order, and the frequency with which they were exchanged must come as a shock to modern man so pressed for time. The wide distribution of the letters themselves often brought new invitations to correspond, as when the Abbé Nollet suggested to Delisle that he write to Perelli, professor of mathematics at Pisa, concerning Italian observations to be made in concert with La Caille's at the Cape of Good Hope.[79] Invariably, the single interest which brought the original exchange of letters into being and animated its early content was eventually extended to a host of related subjects if the correspondence itself persisted. Thus we find Delisle writing to Eustachio Zanotti at the Institute of Bologna on the quality of English astronomical instruments, on the work of Maire and Boscovich on "la mesure de la meridienne de Rome d'une mer à l'autre," and on a plan for the reciprocal

[78] *ibid.*, p. 368. The italics are mine.
[79] Obs. de Paris, B.1,1-8 (140), No. 15.

exchange of books and pamphlets through Count Lorenzy, the French chargé d'affaires at Florence.[80] Underneath it all, as Delisle wrote to Zanotti in the same letter, lay the desire to see the forward movement in astronomy take place in steps approved of "par la pluralité des suffrages de tous les Astronomes."[81] The slow publication of memoirs and transactions in the eighteenth century, often several years behind schedule, also threw the burden of prompt communication upon the letter writer.

But it is also easy to be overly impressed by the penned interchange of ideas and of opinion in the process of formation. In the long run, the solid foundation upon which the working machinery of astronomy evolved in these years is to be found in the memoirs and notices, great and small, which were published under the auspices of scientific societies, by the patronage of royal friends or by the subsidy of men of wealth. The kind of *Avis aux Astronomes* which La Caille published before his departure for the Cape of Good Hope was just this sort of thing, and we shall see it repeated many times for the transits of Mercury and Venus.

The chance of deducing a satisfactory value for the solar parallax from a transit of Mercury still lay within the range of possibilities, especially since the 1753 transit would occur rather centrally on the solar disk; that is, the chord formed by the path of the planet across the sun's face would be longer than had hitherto been the case—a factor which would tend to make possible greater accuracy of observation as a result of the increased duration of the transit itself. The flurry of excitement created by La Caille's African venture had not yet subsided when discussions on the transit of Mercury began. Indeed, even while some of La Caille's incomplete reports were still reaching Paris, plans for observing the transit of Mercury were already under discussion at the Académie.[82] During most of 1752, the circulation of letters dealing with the transit among astronomers must have been very heavy indeed, if the extant material in Delisle's papers is a fair sample, for during that period nearly

[80] *ibid.*, No. 13.   [81] *ibid.*, fol. 2.
[82] *Proc. Verb.*, 1753, fol. 171.

every letter of his carried some discussion of the approaching event. Delisle was particularly anxious to exploit the method (eventually to be applied to the transits of Venus) which he had begun to develop as early as 1723 for the transit of Mercury of that year.[83] In the set of instructions dispatched to St. Domingue through the Marquis de la Galissonière, he placed great emphasis on determining the exact longitude of the place of observation. If one wishes to be able to determine the solar parallax from the observations of a single contact of the planet with the sun, he insisted, it is as necessary to have a precise value for the longitude of the station as it is to observe accurately the precise moment of contact in the transit. "C'est pourquoi," he writes, "l'on recommande d'y faire avec la plus grande precision possible, toutes les . . . observations ordinaires dont on se sert pour déterminer precisement les Longitudes terrestres."[84]

One aspect of the French attempt to arouse interest in this transit of Mercury for the sake of the solar distance emerges from a Franco-British exchange of letters. In a letter to the secretary of the Royal Society, Delisle points out that the phenomenon will be visible in the English colonies, and that an exact observation of the transit made there in conjunction with those to be obtained from the East Indies would be "d'une extreme Consequence" for the calculation of the solar parallax. Consequently, he begged Birch, as he had Bradley in another letter, "de vouloir bien proposer la chose à la Societé Royale et . . . dc vous employer de tout vôtre Credit pour nous procurer de vos Colonies angloises les observations les plus exactes qu'il sera possible."[85] The irony of the situation is that at about thc very time that Delisle was making this strong appeal to the Royal Society, the English colonists, as we have seen, were helping themselves to his own memoir and, in effect, seeking

[83] See Delisle, "Sur le dernier passage attendu de Mercure. . . ," *Mémoires*, 1723, 105-110. Also, "Observation du passage de Mercure . . . le 9 Novembre 1723 . . . ," *ibid.*, 306-343, and "Extrait d'une lettre de M. Delisle, écrite de Petersbourg le 24 Août 1743 . . . ," *ibid.*, 1743, pp. 419-428.

[84] "Instructions pour l'observation du passage de Mercure sur le Soleil du 6 May 1753 dont la Sortie peut être observée au Cap François dans l'Isle de St. Domingue," Obs. de Paris, Delisle mss, B.1,1-8 (140), No. 19, fol. 2.

[85] Obs. de Paris, B.1,1-8 (140), No. 23, fol. 1. Letter dated Paris, 30 November 1752.

to fulfill that request. This sort of notification was repeated many times over by the Delisle correspondents[86] and no doubt by others echoing their interest.

Some time between 30 November 1752 and 2 March 1753,[87] Delisle produced a printed *avertissement* on the transit of Mercury which was similar in intent to La Caille's *Avis aux Astronomes*, containing general instructions on how to observe the phenomenon. He also produced a drawing representing that part of the Mercury transit which would be visible at Paris (Figure 7). The manner in which these privately printed notices were circulated is discernible from a letter to Wargentin in Stockholm.[88] Delisle sent Wargentin six copies of the *avertissement* to present to his academy and to the astronomers of Sweden, "à fin d'encourager les uns et les autres, à faire tous leurs efforts pour nous procurer des ôbservations exactes de ce passage dans lieux les plus convenables."[89] At the same time he promised to send at a later date a much larger number to be turned over to a bookstore for sale to the general public. This Delisle asks Wargentin to do so that he could at least partly reimburse himself for the expense of publishing the notices, and also to enable him to undertake work of a similar sort in the future.[90] The dispatch of twenty-five copies of the same pamphlet to a friend at The Hague[91] for distribution in that area of Holland makes it possible for us to take notice of what Delambre grudgingly called Delisle's original and important contribution

---

[86] For example, the letters to the Rev. Père Chevalier at Lisbon, *ibid.*, No. 23; George Kraz, Professor of Mathematics at Ingolstadt, *ibid.*, No. 56; Marinoni at Vienna, *ibid.*, No. 57; Wargentin at Stockholm, *ibid.*, No. 59; and Felix Sabatelli, Professor of Mathematics at Naples, *ibid.*, No. 65.

[87] It is possible to establish these dates with some certainty, because in the letter to Birch, written 30 November 1752, he mentions the work as in progress, whereas by 2 March 1753, he is able to enclose six copies of it in a letter to Wargentin. Obs. de Paris, B.1,1-8 (140), Nos. 23, 59.

[88] Wargentin was the Secretary of the Swedish Academy of Science, and we shall see that he played a large role in the transit of Venus observations, as indeed so many of Delisle's correspondents did.

[89] Obs. de Paris, B.1,1-8 (140), No. 59, fol. 1.

[90] *ibid.*, fols. 1-2. The cost of such an operation can be ascertained, in connection with the transit of Venus for which an account exists. I have no idea, however, as to what these pamphlets sold for to the general public.

[91] A Mr. Koenig, Professor of Law in that city, *ibid.*, No. 72, fol. 1. Letter dated Paris, 28 March 1753.

to astronomy. This was the method of handling the calculation of transits by means of heliocentric coordinates and the technique of revealing the places of entry and exit of the planet upon the sun as seen from the earth by means of stereographic projections—projections, that is, of appropriate parts of the earth's surface in a plane to locate the zones of visibility for the phenomenon.[92] Of the heliocentric coordinates it need only be said that they simplified the problem of calculation, whereas the idea of the projection was much more important. Delisle repeated the operation for the transit of Venus in 1761, and Lalande, much influenced by this work of his master, did the same for that of 1769.[93]

A few days after his correspondence had revealed the existence of the geographical projections for the Mercury transit of 1753, the *mappemonde*, Delisle submitted it to the Académie together with the *avertissement* which he had already been privately distributing for some time.[94] In addition to locating the visible regions of the phenomenon, Delisle's *mappemonde* made it possible to determine at a glance the advantages or disadvantages of each station, and by means of circles traced on the map the time at which each observer ought to see the entry or exit of the planet from the solar disk. It would not be germane to this study to pursue in detail the Mercury transit of 1753, save to underscore its role in the preparations for Venus in 1761 and 1769, and to reveal that the event itself introduced into the astronomical world, or brought forward for the first time, many of the principals who were to take part in the transit of Venus expeditions to all parts of the world. Le Gentil, for example, who was to go to India, Chappe d'Auteroche to Siberia in 1761 and California in 1769, and Pingré to the Isle Rodrigue off Madagascar were to make their mark with the advent of this Mercury transit.[95]

[92] Delambre, *op.cit.*, pp. 325-327.
[93] Lagrange eventually submitted this geometrical technique to analytical treatment, but though he demonstrated the feasibility of this more sophisticated handling of the problem, it was really no simplification over the geometric method, *ibid.*, p. 327.
[94] *Proc. Verb.*, 1753, fol. 171.
[95] Le Gentil and Chappe d'Auteroche observed the event from the inner terrace of the Observatoire de Paris, *Proc. Verb.*, 1753, fols. 173-179. Note that

The report which Le Gentil read to the Académie on the transit of Mercury took place in the same year that he became a member of that august body of savants.[96] Like so many scientists of the eighteenth century, Guillaume-Joseph-Hyacinthe-Jean-Baptiste le Gentil de la Galaisière originally planned to enter the Church. Later, in the early nineteenth century, Cassini IV explained Le Gentil's attraction toward an ecclesiastical career on the grounds that "il était toujours sûr alors de s'attirer un certain respect et des égards qui le plaçaient souvent au-dessus de son rang et de sa fortune," adding the quip that Le Gentil "ne garda l'habit d'abbé que jusqu'au moment ou le titre de savant lui procura une considération et une existence moins équivoques."[97] But the curiosity which one day took Le Gentil to the Collége Royal to hear Delisle lecture was great enough to sustain his interest in astronomy and he came regularly. Delisle soon recognized his worth and presented him to the seventy-one-year-old Jacques Cassini, *doyen des astronomes de l'Académie*, who invited him to work at the royal observatory under the supervision of his son, Cassini de Thury, and his nephew Maraldi.[98]

By 1753, Le Gentil was therefore a well-trained observer contributing regularly to the memoirs and meetings of the Académie. His explanation of his inability to determine the internal contact of Mercury in this transit was thus of some importance. The speed of the planet, he felt, made it impossible to obtain an

---

the report which Le Gentil read to the Académie was misplaced in the *Procès-Verbaux*. The Mercury transit took place on 6 May 1753 so that no observational report of it could possibly have taken place before that date. The transcription of Le Gentil's statement has been accidentally interchanged with one for 30 June 1753, which is when he actually delivered it.

The transit of Mercury also brought Pingré into contact with Delisle in a discussion of calculations concerning it. Cf. Obs. de Paris, B.1,1-8 (140), No. 77. His report on the observation itself, made together with Le Cat and others at St. Lô was read to the Académie on 16 May 1753, by Le Monnier, *Proc. Verb.*, 1753, fols. 355ff. There seemed to be enough confusion caused by a curious crowd, the wind, and a multiplicity of observers to lead Pingré to comment that "l'observation . . . fut ce jour là une vraye tour de Babel. . . ." *ibid.*, fol. 358.

[96] Cassini, J.-D., *Éloge de M. Le Gentil, membre de l'Académie royale des sciences de Paris* (Paris: D. Colas, 1810), p. 8.

[97] *ibid.*, p. 6.

[98] *ibid.*, p. 7. Cf. also "Legentil de la Galaisière," *Biographie Universelle*, XXIII, 618.

accuracy greater than two seconds.[99] Le Gentil was therefore one of the first after 1753 to depreciate the value of transits of Mercury for obtaining the solar parallax. Incidentally, in reporting to the Académie the observations which he had made together with Le Monnier and La Condamine at the Château de Meudon, Lalande pointed out that all the tables had been inaccurate, "les meilleurs tables differeoient prodigieusement entre elles"; that those of De la Hire gave the exit time 8 hours earlier than those of Halley, and Delisle's, which were the best of all, gave it 17 minutes too late.[100] The degree of predictability based on the available astronomical tables only a few years before the transits of Venus therefore seemed very unsatisfactory indeed.

With the transit of Mercury behind them and the recognition that its results had not fulfilled the expectations held by some on its behalf, the membership of the Académie turned its attention to the transits of Venus of the coming decade. The half century between Halley's now well-publicized insight and the period of its fulfillment had gone by. It was time to actively consider the transits of Venus.

Addressing himself to this particular problem in the light of the failures connected with Mercury, and equally aware that the published tables of Halley, those of 1749, were imperfect enough to have required several corrections and improvements, Le Gentil undertook to reconsider the problem of determining just when, where, and for how long Venus would be seen upon the face of the sun at transit time. Obviously designed to make possible a successful observation of the transit of Venus, the memoir which he produced and read to the Académie on 30 June 1753[101] also served to rekindle the contemporary imagination. The principal elements of the theory of Venus are based upon observations of the planet in conjunction with the sun. Since this

[99] *Proc. Verb.*, 1753, fol. 176.
[100] *ibid.*, fol. 435.
[101] *ibid.*, fols. 450*-456*. The stars indicate that the pages are incorrectly numbered, the starred numbers acting as numbers added after the pagination has been established. This memoir has also been printed. Cf. *Mémoires* (Amsterdam: 1762), 1753, pp. 43-55. Also see the discussion of it in *Histoire*, 1753, pp. 326-330.

was also a rather rare event, the data required to build up a complete theory of the planet was still rather limited. Halley's work in this respect was based on the Rudolphine Tables, and when Jacques Cassini constructed his new tables in the last years of the seventeenth century, he discovered errors in the Rudolphine Tables of direct importance to the theory of Venus. Furthermore, Cassini's new values had the good fortune to be closely corroborated by the inferior conjunction of Venus with the sun in 1737.[102] Consequently, since the circumstances of this conjunction were fairly close to those expected in 1761 and 1769, Cassini calculated the various elements of the transits for that period. Needless to say his results differed from those of Halley, and it was the existence of these two sets of contradictory data that precipitated Le Gentil's labors in this particular vineyard. His correction favored Cassini's data. In fact, two corrections in Halley's work were required: one in the expected duration for 1761, and the other with regard to its visibility at Paris in 1769, to which Le Gentil gave an affirmative opinion against Halley.[103] The method which Le Gentil used to re-examine the Halley and Cassini calculations was the one published by Delisle in connection with the Mercury transit of 1723.[104] In addition to the usual peroration calling for the maximum number of observers of the transit of Venus from widely separated points on the earth, Le Gentil added a drawing which represented the chords to be described by Venus in 1761 and 1769 following the calculations of various earlier scholars (Figure 9).[105]

In the period which followed Le Gentil's memoir, dissatisfaction with Halley's tables continued to bring forth a steady stream of constructive criticism. Though not all of it was per-

[102] *ibid.*, fol. 451*. These observations upon which Cassini built his theory of Venus were discussed in his *Elémenes d'Astronomie* published in Paris, 1740.

[103] *ibid.*, fols. 452*-453*. Halley had believed that the entire transit of 1769 would not be seen at Paris, but Le Gentil demonstrated that the entry of Venus onto the sun would.

[104] *ibid.*, fol. 453.

[105] This diagram is taken from a manuscript copy in the Delisle papers, Obs. de Paris, A.6,9 (61,16,c). It seems likely that this was attached to a copy of Le Gentil's memoir which was sent to Delisle. There is no diagram accompanying this memoir in the *Procès-Verbaux*. Compare this with Whiston's representation thirty years earlier.

tinent to the transit of Venus,[106] the fact that inadequacies in the tables were frequently revealed meant, at the very least, that they must always be used with a large measure of caution. It ought to be said at this point that the frequent reiteration of Halley's shortcomings in this study is by no means intended to reduce his greatness. Although there were errors in his tables directly due to his own miscalculations and these errors did bear upon the transits of Venus, as we have already seen from Le Gentil's memoir alone, some were a result of the data to which he fell heir in the Rudolphine Tables, and others were simply exposed by the growth in astronomical information after his death.

In 1759 Lalande requested the Académie to appoint a committee to prepare a new edition of Halley's tables on planets and comets, to incorporate the corrections made to date, and to add supplementary material on the satellites of Jupiter and the fixed stars. And to this task Clairaut and La Caille were assigned.[107] Maximilian Hell, about whose activities in connection with the transits a long controversy was to persist down to our own day,[108] was elected a correspondent of the Académie in this period and assigned to La Caille.[109] Indeed the same span of time brought two other principals in the future transit of Venus expeditions into the Académie des sciences, Alexandre-Gui Pingré in 1756,[110] and Jean Chappe d'Auteroche in 1759.[111]

During the two- to three-year interval immediately preceding the transit of Venus in 1761 discussion of specific details concerning its observation was most 'intense. The location of stations, the times of contact, and the anticipated duration of the event were vigorously debated in the halls of the learned academies and on the pages of their journals. A mere glance at

[106] For example, see de Vaussenville's correction of Halley's tables in connection with the timing of the solar eclipse of 1753, a correction which was sustained by Le Monnier and Maraldi, Proc. Verb., 1754, fols. 126, 139ff.

[107] Proc. Verb., 1759, fol. 462.

[108] George Sarton, "Vindication of Father Hell," Isis, xxxv, 2 (1944), 97-105. This should be the last word on the Hell controversy, to be discussed below with biographical details as well.

[109] Proc. Verb., 1758, fols. 957-958.

[110] ibid., 1756, fol. 399. Biographical sketches of Pingré and Chappe will be given in connection with their voyages.

[111] "Éloge de M. l'Abbé Chappe," Histoire, 1769, p. 164.

a bibliographical chronology of the transit will reveal the plethora of publications which came into being at this time. Essays, pamphlets, notices, lectures—in short, every form of public address was used to call attention to the importance of the approaching transit of Venus. Their common denominator was more often than not an appeal to patriotism and to the ideals of the Enlightenment. But sometimes these paeans turned into personal polemics and reached the popular magazines and *feuilles volantes* like *L'Avantcoureur, La Feuille nécessaire,* or *Le Censeur hebdomadaire.* On such an occasion, the controversy that raged through the articles and letters-to-the-editors of these journals also played its part in publicizing the transit of Venus. The mere fact that the editors of these publications saw fit to print the entire correspondence in such a case is also a measure of the public interest in the problem. One such incident worth noting concerned Halley, Delisle, the choice of an observational station in North America, and a certain Trébuchet, sometime officer to the Queen, maternal grandfather of Victor Hugo, astronomical amateur, and recently Delisle's assistant.[112]

Having taken it upon himself to recalculate the elements of the transit of Mercury in 1753[113] and to produce the special memoir and *mappemonde* to bring its importance to the attention of the scientific world, it was quite natural for Delisle to do the same for the far more important transit of Venus in 1761. For the successful accomplishment of this purpose the cautious use of Halley's specific predictions was now a *sine qua non* of the calculator. With this consideration in mind, and following the method which he had brought to a high level of development by successive application to the transits of Mercury in

[112] Trébuchet's portrait is to be found in the Victor Hugo house, in the lovely Place des Vosges (Paris), as is the information contained in this statement. It is the portrait of a smug, self-centered individual, but it does reveal his interests, for he is seen with octant in hand and a chart before him.

[113] The Mercury transit of 1756, while without the interest of its immediate predecessor, precipitated another memoir from Delisle, part of which dealt with the transit of Venus. "Observations du passage de Mercure sur le disque du Soleil, le 6 Novembre 1756; avec des réflexions qui peuvent servir á perfectionner les calculs de ces passages, & les élémens de la théorie de Mercure déduites des observations," *Mémoires,* 1758, pp. 134-154. The transit of Venus reference is to be found on p. 140.

1723, 1743, and 1753, Delisle re-established the elements of the transit of Venus. In this task he was assisted by Trébuchet and Libour, the latter a *Maître de Mathématique* who had been associated with Delisle as assistant in computation for about twelve years.[114] By November of 1759 Delisle's preliminary results could be read to the Académie, and on the twenty-first of that month he reported "sur le passage de Vénus sur le Soleil qui doit arriver le 6 juin 1761."[115]

The main purpose of this early report was to reveal a basic error in Halley's choice of stations and in his estimation of the duration of Venus upon the sun. With the center of the earth as a theoretical observational station, Halley had indicated that the difference in the duration of the transit between it and a station in the East Indies would be 12 minutes; that is, the transit would last 12 minutes longer when seen from the East Indies location than when theoretically seen from the earth's center. For his northern station Halley had selected Port Nelson on Hudson's Bay, and he foresaw an increase of about 6 minutes in duration at this location as compared to the earth's center. Consequently, between a station in the East Indies and one at Port Nelson, a total difference of about 18 minutes in time was to be expected, enough, in other words, to make these locations important Halleyan stations. However, the phenomenon itself at Port Nelson would not be visible under the best possible conditions. The entrance of the planet upon the sun's face would take place in the evening before sunset, and exit would occur the next morning a little after sunrise. But Delisle's calculations predicted another state of affairs entirely. Choosing Pondichery as his East Indian location, he found an 8-minute diminution of the transit at that site in comparison with the earth's center, rather than a 12-minute prolongation. At Port Nelson, Delisle discovered that the 6-minute interval in question would also be one of diminution, so that the total difference in transit time to be expected between Pondichery and Port Nelson would be about 2 minutes, rather than Hal-

[114] From a manuscript copy of a letter by Libour to *La Feuille nécessaire, contenant divers détails sur les sciences, les lettres & les arts* (hereafter cited: *La Feuille*), dated "à Paris le 30 dec. 1759." Obs. de Paris, A.3,12. (34.b.6), fol. 1.
[115] Proc. Verb., 1759, fol. 771.

ley's 18. This considerably reduced the importance of the Port Nelson station, and Delisle begged his colleagues to interest themselves in this problem by examining both his method and his calculations. At this meeting Delisle also suggested that consideration of the Port Nelson station should be abandoned, and that a station farther north should be selected, preferably at a latitude of 65 degrees.[116]

Shortly after this memoir was read to the Académie, a report of its conclusions reached the general public, under the title "Découverte sur le passage de Vénus au-devant du Soleil, attendu le 6 Juin 1761," in one of the popular weeklies referred to earlier.[117] It stressed the importance of the transit of Venus, gave Halley's comments and conclusions on the subject, and then indicated that Delisle, "qui avoit eu l'avantage de prédire le célèbre passage de Mercure . . . plus près de son calcul que de celui d'aucun autre Astronome," had discovered that there were errors in Halley's work. The problem of choosing the stations therefore remained to be solved, and toward this end, *La Feuille nécessaire* informed its public that Delisle had been preparing and was about to publish a *mappemonde* which would reveal this for every possible useful location. A memoir explaining its use would accompany the *mappemonde*.[118] But this useful and mild-mannered announcement was not everywhere accepted in the spirit in which it was offered.

The contentious Trébuchet found much to displease him in this article, and in the next issue of *La Feuille nécessaire* he quickly revealed his mood with the opening underscored phrase, "*A chacun le sien, c'est justice.*"[119] What had apparently aroused

---

[116] *ibid.*, fols. 771-772. Port Nelson (today in the Province of Manitoba) is very roughly at 58 degrees north latitude and 92 degrees west of Greenwich. Delisle's 65 degrees north latitude would mean a station a little below the Arctic Circle.

[117] *La Feuille*, No. 45 (17 Décembre 1759), 712-715.

[118] *ibid.*, p. 714. This publication makes the statement that Delisle has already shown the *mappemonde* at the Académie and that, in fact, Le Gentil has already volunteered to go to Pondichery. The record seems to indicate otherwise, for the *Procès-Verbaux* shows that Delisle first presented the *mappemonde* to the Académie on 30 April 1760; and the first evidence of Le Gentil's plans for departure appears in the *Procès-Verbaux* on 19 January 1760. Of course it is quite likely that both were discussed before any record was made, and in twentieth-century fashion, the information leaked to the press.

[119] "Extrait d'une lettre de M. Trébuchet, ancien officier de la Reine, du 18 Décembre 1759," *La Feuille*, No. 46 (24 Décembre 1759), p. 724.

his wrath was the honor given to Delisle for the discovery of
Halley's error. He objects to the announcement made "sous
le titre pompeux de *Découverte*," but quickly wishes to claim
that honor for himself.[120] Then, as if to strike back at Delisle
for this alleged impropriety, Trébuchet goes on to say that
there will be no need for this *mappemonde*, which, though
likely to be quite well done, will be confusing and useless, or
at best redundant, for it will only repeat what he himself is
about to proclaim: that the stations of observation ought to be
on "la côte occidentale de la nouvelle Hollande d'une part, &
de l'autre, les rives du Gange, Pekin & Torneo."[121] The claims
were thus advanced and the fruitless argument for priority of
discovery or invention joined. Before the month was over, Libour
wrote the editors of the same journal to inform them of the
facts "mieux que Mr. Trébuchet mesme." His twelve-year-old
association with Delisle from student days at the Collège Royal
to the moment in question, his role as computor in the preparation of the *avertissement* and the *mappemonde* on the 1753
transit of Mercury, and finally his assertion that both his work
and Trébuchet's were constantly under the guidance, supervision, and frequent check of Delisle, were all marshaled by Libour
to reject Trébuchet's argument against Delisle. With the hope
of sealing the argument, he indicated that the *mappemonde*
would reveal more than Trébuchet had indicated, that in fact
the very places he had named were those for which he was
asked to make the appropriate calculations by Delisle, and finally, that a memoir which he himself was preparing on the
transits of Venus, designed to be the definitive statement on
the astronomical elements involved, was wholly due to Delisle.[122]

The *mappemonde* and its accompanying memoir were finally
presented to the Académie on 30 April 1760[123] and shortly thereafter approved by the committee appointed to examine them.[124]
Within a week they were published and an announcement to
that effect, together with a brief but complimentary description

120 *ibid.*, p. 725.
121 *ibid.*, p. 726.
122 Obs. de Paris, A.3,12 (4.b.6), fols. 1-3.
123 *Proc. Verb.*, 1760, fol. 257.
124 *ibid.*, fol. 266. The committee consisted of Le Monnier and Pingré.

was carried in *L'Avantcoureur*.[125] The same memoir which De-lisle had presented to the Académie was later reprinted in the July issue of the *Journal des Sçavans*, for the same year, obvi-ously because of its importance and the favorable manner in which it had been received. In spite of all this evidence in ap-proval of Delisle's work, Trébuchet's arguments were not to be stilled; indeed, even in the midst of this hostile environment, they had persisted in the pages of the *Mercure de France*,[126] *L'Avantcoureur*,[127] *Le Censeur hebdomadaire*,[128] and, after the publication of the *mappemonde*, in the *Journal des Sçavans* as well.[129] In themselves, the long exchange of letters full of petty quarrels and personal animosity came to signify nothing. Their publication, however, did serve to reveal that the interest of the literate public in transit affairs was great enough to warrant their discussion in the pages of the popular journals.

In the midst of this debate issues of greater importance were being resolved. At the meeting of the Académic for 19 January 1760, Le Gentil announced that he was about to leave for Pon-dichery to observe the transit there. The arrangements for this undertaking, though they must have been discussed among the membership, are not to be found in the *Procès-Verbaux* of the Académie. Indeed, Le Gentil's announcement even antedates Delisle's presentation of the *mappemonde* to the Académie and, therefore, the prolonged discussion of the transits which it pre-

[125] In the section "Astronomie," *L'Avantcoureur*, No. 19 (26 Mai 1760), 297-298.

[126] "Mémoire sur l'éclipse du Soleil par Vénus du 6 Juin 1761," *Mercure de France* (Janvier 1760). A manuscript copy of this article is in Obs. de Paris, A.3,12 (34.b.7). The same article was later printed as a letter under the title "Lettre de M. Trébuchet, ancien officier de la maison de la Reine, a Messieurs les auteurs du Journal des Sçavans, sur l'éclipse du Soleil par Vénus, du 6 Juin 1761," *Journal des Sçavans* (Novembre 1760), pp. 733-737.

[127] "Extrait d'une lettre de M. Trébuchet d'Auxerre du 3 Juin 1760," *L'Avant-coureur*, No. 21 (9 Juin 1760), 350-351.

[128] "Lettre au R. R. P. P. journalistes de Trévoux," *Le Censeur hebdomadaire*, Tome III, Article XXXVI (1760), 306-309. "Lettre de M. Trébuchet a M. d'Aquin, au sujet du passage de Vénus de 1769," *ibid.*, Tome III, Article XLV (1760), 385-396; "Lettre de M. Trébuchet . . . ," *ibid.*, Tome II, Article LIX (1761), 369-378; "Suite de la lettre de M. Trébuchet. . . ," *ibid.*, Tome III, Article I (1761), 3-14.

[129] "Lettre de M. Trébuchet . . . en reponse a celle d'un Académicien, inserée dans le Journal d'Avril 1761, au sujet des calculs fait par M. Delisle. . . ," *Journal des Sçavans* (Fevrier 1762), pp. 101-109.

cipitated. Only the barest outline of what must have been rather involved political and economic negotiations is discernible. Acting on his own, Le Gentil apparently approached the Duc de Chaulnes with his plans for the observations, and was referred in turn to the Comte de Saint-Florentin, Secretary of State and honorary member of the Académie, and M. de Silhouette, Controller-General, both of whom, to their eternal credit, approved the project in spite of the war with Britain, then well under way. The cooperation of the Compagnie des Indes was obtained through the Secretary of State and a M. Boutin, its "Maître des Requêtes, & Commissaire pour le Roi,"[130] though in what manner and to what extent remains unknown.[131] Undoubtedly between 19 January 1760, when Le Gentil's name last appeared on the list of those present at the meeting of the Académie, and 26 March when he sailed from Brest,[132] he was preparing for the long voyage, but the full nature of those preparations is not revealed, save for some brief information on the instruments he took with him. To conclude the bare outline of his preparations, we know that his final orders were delivered to him by the Duc de la Vrillière, that before boarding ship he was given the choice of sailing on the sixty-four-gun *Comte d'Artois* or the fifty-gun *le Berryer* and that he chose the latter.[133]

The memoir on the transits which Le Gentil published in the *Journal des Sçavans* on the eve of his departure, in addition to supplying most of the above details, was a general review of the transit problem, historically presented and scientifically analyzed. Especially concerned with the weakness of the Port Nelson station at 56 degrees north latitude, on the basis of both Cassini's tables and the corrected posthumous Halley tables, Le Gentil supported Delisle's suggestion that a station farther north

[130] Le Gentil, "Mémoire de M. Le Gentil au sujet de l'observation qu'il va faire, par ordre du Roi, dans les Indes Orientales, du prochain passage de Vénus pardevant le Soleil," *Journal des Sçavans* (Mars 1760), 139.

[131] It will perhaps be possible to project the experience of the British scientists in this respect with regard to the East India Company in order to more fully understand the organizational procedure of such a scientific expedition.

[132] "Le Gentil de la Galaisière," *Biographie Universelle*, XXIII, 618.

[133] Le Gentil, *Voyage dans les mers d'Inde (1760-1771), fait par ordre du Roi, à l'occasion du passage de Vénus, sur le disque du Soleil, le 6 Juin 1761 & le 3 du même mois 1769* (Paris: Imprimerie royale, 1779-1781), I, 1-2.

was needed, but in view of obvious difficulties in fulfilling that need, offered alternatives. In his own words:

Il faudroit donc (si on vouloit faire l'observation sur la côte Orientale de la Baye d'*Hudson*) choiser un endroit plus au Nord de quelques degrés que n'est le Port *Nelson*; mais il ne sera pas aisé à tout Astronome d'aller s'établir dans la Baye d'*Hudson*, à moins que les Anglois ne prennent le soin de s'en charger; il faudra . . . se contenter des observations que l'on va faire dans les *Indes*, & de celles qu'on pourra faire en *Europe*, soit à *Stokolm*, soit à *Petersbourg* . . . soit à Londres, soit à Paris, soit enfin à Bologne, en Italie, lieux qui ne Manqueront pas d'habiles Observateurs, quoique moins avantageusement placés qu'à la Baye d'Hudson.[134]

Thus the process of eliminating Port Nelson as an observational station in 1761 was strengthened by Le Gentil's analysis, and the criticism which was voiced in France saved the English much useless effort. Le Gentil then concluded his study with the patriotic apostrophe that it was France alone which was undertaking "les grandes entreprises qui concourent si fort au progrés des Sciences les plus utiles, l'Astronomie, La Géographie, & la Navigation."[135]

The unsung embarkation of Le Gentil for the orient, in an enterprise which was to demand so much—more than ten years of his life—and yield so little—he was never to see a transit of Venus—was followed by an acceleration in the pace of all activities connected with the transits. At the 19 April meeting of the Académie, La Caille read a letter from the secretary of the Russian Academy at St. Petersburg asking "si quelqu'un des astronômes de L'Académie voudroit aller en Sibérie observer le passage de Vénus," under the auspices of the Empress.[136] Both Pingré and Chappe volunteered for this task, partly out of interest in the subject and partly because they were the only astronomers in the Académie who had not yet made any astronomical voyages. Between them they decided subsequently that Chappe would go to Russia, there to be seconded by the St.

[134] Le Gentil, "Mémoire . . . au sujet . . . du prochain passage de Vénus . . . ," *Journal des Sçavans* (Mars 1760), p. 137.
[135] *ibid.*, p. 139.
[136] *Proc. Verb.*, 1760, fol. 239.

Petersburg astronomers, and that Pingré would go on any other projected voyage which might come up.[137]

The *mappemonde* which Delisle presented was similar in purpose and construction to the one drawn for the transit of Mercury in 1753, and all the richer for that previous experience. It consisted of two hemispheric projections upon which various curves relating to specific elements of the transit of Venus were drawn with a high degree of precision.[138] In each hemisphere the curves divided the map into three regions, each representing that part of the earth illuminated by the sun at the moment of Venus' entry upon its disk, during the full time of its transit across the disk, and at the moment of exit from it.[139] Additional curves were given other designations: for entry there were subdivisions to indicate those areas where entry would take place at sunset, and those where it would occur at sunrise; and the same thing was done for the moment of exit. In this manner every zone and degree of visibility of the transit for the entire globe could easily be distinguished. To improve the utility of the *mappemonde*, the areas in which the entry could be seen were colored blue, those of total duration red, and those of exit yellow.[140] The existence of a normal map beneath these lines and colors therefore permitted the immediate identification of those places likely to serve as stations of observation, not only because of the characteristics of the phenomenon itself, but also because of the geographical position of the station, its relative accessibility, and its proximity to civilization. Further refinements were added to the *mappemonde* in the form of curves

[137] "Du Passage de Vénus sur le Soleil; Annoncé pour l'année 1761," *Histoire*, 1757, p. 77.

[138] The description of Delisle's *mappemonde* is based entirely on textual sources, for I have never seen the drawing itself. In spite of the apparently large number of copies which were made, none exist in the Delisle archives or in those of other individuals connected with the transits. However, other *mappemondes* were produced in imitation. I have seen those and shall have occasion to refer to them.

[139] Delisle, "La Description et l'usage de la mappemonde dressé pour le passage de Vénus sur le disque du Soleil qui est attendu le 6 Juin," Obs. de Paris, A.3,12 (34.b.34), fol. 2.

[140] *ibid.*, fol. 3.

designating the times of contact between Venus and the sun for various areas, counting time from the Paris meridian.[141]

With the *mappemonde* as a foundation, it was therefore easy to choose the stations of observation. For example, Delisle indicated that of all places where the total duration of Venus would be both visible and accessible, Tobolsk in Siberia would have the shortest apparent duration and Batavia in the Dutch East Indies the longest.[142] An examination of Figure 8, which is an inadequate Dutch derivation from Delisle's map, will give the relative separation of these two places. At the same time, some idea of the original *mappemonde* can be obtained from this Dutch copy.[143]

As might be expected, the memoir emphasized the importance of stations at which either entry or exit could be observed as well as those at which the total duration would be visible, for this was the very feature in which Delisle's method differed from Halley's. Stations were chosen to fulfill this condition at Yakoutsk and Kamchatka on the one hand, and the Cape of Good Hope and St. Helena on the other. Additional possibilities selected as Delislean stations were Peking, Macao, Archangel, Torneo, and St. Petersburg,[144] and these would be of great advantage in connection with the aforementioned stations. In view of the public debate with Trébuchet referred to earlier, it is proper to note that throughout this memoir Delisle emphasized in detail the important work done by Libour in supplying the calculations which were the foundation of the *mappemonde*.[145] As he had done in 1753, Delisle again called upon the astronomers who were to take part in this great venture to pay special attention to the length and quality of the telescopes used and above all to the effect of smoked and colored lenses when observing against the background of the sun.[146]

---

[141] *ibid.*, fols. 4-6.   [142] *ibid.*, fol. 8.

[143] Dirk Klinkenberg, *Verhandeling, Beneffens de Naauwkeurige Algemeene en Byzondere Afbeeldingen van den Overgang der Planet Venus voorby de Zon, en den 6 Juny 1761 des Morgens* (The Hague: no publisher, 1760), p. 41.

[144] Delisle, "La Description et l'usage de la mappemonde. . . ," *op.cit.*, fols. 9-10.

[145] *ibid.*, fols. 14-17. Libour later read a separate memoir to the Académie on the subject of the calculations. *Proc. Verb.*, 1760, fol. 298.

[146] *ibid.*, fol. 18.

One interesting aspect of Delisle's *mappemonde* and the *avertissement* that went with it is the question of its distribution. It is virtually impossible to determine the extent of the sale of the two publications, but fortunately Delisle has left us a kind of mailing list of those who were to receive them without charge. The distribution was virtually as wide as the civilized world and came to two hundred copies,[147] which probably cost Delisle thirteen pounds.[148] Because of the rare importance of such a list in revealing the names, places, and even the manner of distribution of scientific material in the eighteenth century, it will be given in an appendix.

With Delisle's *mappemonde* available and its world-wide distribution under way, and with one of its own members already embarked for the Orient, the Académie finally undertook a full discussion of the transits of Venus—a discussion precipitated in part by the force of these events and in part by a memoir which Lalande read at the meetings of 7 and 10 May.[149] Lalande's memoir once again discussed Halley's errors as Delisle had indicated them, and in addition gave a fuller explanation of the geometric and geographic problem in constructing the *mappemonde*, for which purpose he drew one of his own. Additional stations were also suggested by Lalande on the west coast of Africa from the Cape of Good Hope up through the Portuguese colonies of Luanda and Benguela (St. Helena is about 20 degrees west of Benguela in the south Atlantic Ocean). Expressing ignorance as to whether the English would observe at St. Helena, he stressed the value of making certain by sending French observers to Africa.[150] While he was suggesting this plan, he sought to reduce somewhat the degree of precision to be expected from the transit of Venus observations on the basis of the transit of Mercury experience in 1753. Taking five of the observations of interior contact which were made in the

---

[147] Obs. de Paris, A.3,12 (34.b.119) and (34.b.120).

[148] *ibid.*, (34.b.123).

[149] Lalande, "Mémoire sur les passages de Vénus devant le disque du Soleil, en 1761 et 1769. . . ," *Mémoires*, 1757, pp. 232-250. The date according to the printed version is 14 May 1760, but the *Procès-Verbaux* indicate that it was read on 7 and 10 May, and I have taken the latter dates as more reliable because they were recorded closer to the events.

[150] *ibid.*, p. 247.

Paris area by six competent astronomers (including himself), he noted a variation of 5 seconds in the result.[151] But this was only a precaution, for he went on to say that: "L'occasion que nous présente ce célèbre phénomène, est un de ces momens précieux, dont l'avantage, si nous le laissons échapper, ne sauroit être ensuite compensé, ni par les efforts de génie, ni par la constance des travaux, ni par la magnificence des plus grandes Rois; moment que le siècle passé nous envioit, & que seroit dans l'avenir, j'ose le dire, une injure à la mémoire de ceux qui l'auroient négligé."[152]

With the force of this closing argument ringing in their ears, the Académie immediately decided "que Mrs. les astronômes s'assembleroient Mercredi prochain après l'académie, pour délibérer sur ce suject."[153] Unfortunately, the deliberations which took place that day, since they were held after the normal meeting, were never entered into the *Procès-Verbaux*; but by 13 August we learn that Malesherbes, President of the Académie, has spoken to the Comte de St. Florentin, Secretary of State, about the projected voyages. In turn, St. Florentin asked the Académie to draw up a formal memoir on the subject, and Chabert and La Caille were commissioned to write it.[154] Long before he read the joint report to the Académie, Chabert produced a brief memoir of his own on the advantages to be gained from observations made on Cyprus and "quelques Isles de la mer du Sud," in which he offered to go to Cyprus, if arrangements could be made with the Spanish to send observers to the other islands; but nothing ever came of this particular plan for the transit of 1761.[155]

The results of the discussion on the transit of Venus, for which no other record exists, are contained in the report which Chabert and La Caille prepared for the king's minister and read to the Académie on 20 August 1760.[156] Designed to explain "la

---

[151] *ibid.*, p. 248.   [152] *ibid.*, p. 250.
[153] *Proc. Verb.*, 1760, fol. 263.   [154] *ibid.*, fol. 412.
[155] Joseph-Bernard Chabert, "Mémoire sur l'avantage de la position de quelques isles de la mer du Sud, pour l'observation de l'entrée de Vénus devant le Soleil, qui doit arriver le 6 Juin 1761," *Mémoires*, 1757, pp. 49-51.
[156] *Proc. Verb.*, 1760, fols. 421-427. Cf. Chabert, "Mémoire sur la nécessité, les avantages, les objets & les moyens d'exécution du voyage que l'Académie propose de faire entreprendre à M. Pingré dans la partie occidentale & méridionale de

necessité, les avantages, les objets, les moyens d'Exécution et la dépense" of the voyage in question, the very beginning of the report reveals that much has been decided between the time when it was originally requested and this moment of its delivery. With Le Gentil and Chappe now committed, the deliberation had obviously brought about a decision for another voyage to be undertaken by Pingré to the southwestern shores of Africa, in keeping with the emphasis which Lalande's memoir had placed upon this region. The commissioners, La Caille and Chabert, in making the report already had Pingré's approval and, more than that, had had his cooperation in drawing up the document itself.[157]

After reviewing the importance of the solar parallax to astronomy, the nature of the observations, and the possible range of geographical stations, the memoir centered on Tobolsk in Siberia and the as-yet-unknown station in southwest Africa as the most important of all for observing the exit of Venus from the sun's disk, and hence for employing Delisle's method. In all of the southwest African coast above the Cape of Good Hope not one place was known (at least to the French) whose longitude had been precisely determined, in spite of the existence of numerous Portuguese and Dutch trading colonies from Angola to the Gold Coast. Since Delisle's method also required the precise determination of longitude at the observational station, the choice of ports like Benguela or Luanda would serve the double function of improving navigation while advancing astronomy. Furthermore, since the Portuguese had "des établissements solides" there, Pingré would not be without too many of the amenities of civilization. But the question of transportation imposed an unusual limitation on the choice of a Portuguese establishment. Inquiry into the pattern of shipping in and out of these colonies revealed that traffic lanes led to and from Brazil and only indirectly to Europe. A traveler going to these colonies would therefore have to lay over in Brazil for one of the very infrequent Africa-bound ships and repeat the triangu-

---

L'Afrique, à l'occasion du passage de Vénus. . . ." *Mémoires*, 1757, pp. 43-49. This is a briefer printed version which does not mention the expenses involved.
   [157] *ibid.*, fol. 421.

lar route on return. In the eighteenth century this meant years! Consequently, the Dutch colonies on the Gold Coast, where the African trade was heaviest and therefore the shipping most frequent, were deemed the more likely choice. For that matter the two small islands, Prince's Island and St. Thomas Island in the Gulf of Guinea, would also do, and the exact determination of their longitude would be of great value to navigation.[158]

The report also contained an extensive analysis of the climate to be expected in all areas under consideration, with the conclusion that it was very dangerous for all foreigners. In this respect the fear of possible illness brought about the suggestion that Pingré be accompanied by another, lesser ranking astronomer, who might also assist in the investigation or collection of new flora and fauna. Mention was made with approval of the recent news that the British were planning expeditions to St. Helena and Bencoolen on Sumatra. Observations from St. Helena in concert with Chappe's in Siberia would be in keeping with Delisle's method and those from Bencoolen would serve the Halleyan technique. Finally without attempting to cushion the shock of so great an expense, the commissioners simply said that this single expedition would cost about £12,000. A little cold comfort was added to the statement by the assurance to the Minister of State that Pingré would employ "avec toute l'aeconomie [sic] possible les fonds qui lui auront été confiés."[159]

Although La Caille and Chabert thus gave their support to the selection of an African station and things had gone far enough so that letters to the Dutch and the Portuguese for permission to observe were already underway, the issue was far from decided. Sometime between the presentation of the memoir for St. Florentin in August and an additional report read to the Académie by Mairan in November, an important change was made. Pingré was now to go to the Isle Rodrigue[160] in the Mascarenes (a small group of islands east of Madagascar in the Indian Ocean). Whether or not the change was made to safeguard Pingré from the host of potential diseases he would have faced, because of the inclement weather on the African coast

---

[158] *ibid.*, fols. 422-423.   [159] *ibid.*, fol. 427.
[160] *ibid.*, fol. 471.

in June, or possibly because of unfavorable replies from the Dutch and Portuguese governments, we do not know. It certainly was not because Rodrigue would be a better station from which to observe the transit, although it was not an unfavorable one. Indeed, there is some reason to believe in the noncooperative attitudes of the Portuguese or the Dutch and possibly both, for the *Histoire* of the Académie for the time in question implies this in its explanation of the change.

Les difficultés qu'on éprouve dans un pays étranger, influent presque toûjours sur la nature des travaux qu'on y exécute, & toutes choses égales, un François doit souhaiter de pouvoir observer dans des établissemens François, ou l'autorité royale appuie & soutient ses entreprises, ou rien ne peut lui manquer de tout ce qui contribue au succès de ses recherches. . . .[161]

Accordingly, the Académie chose Isle Rodrigue, then in French hands. Perhaps the tone here should not be misconstrued, because the reasons given in addition to the above in support of the change constituted a sound motive. Reports of the weather at the island indicated beautiful skies in June, and since the island was one which it was necessary to be cognizant of on voyages to India, it therefore deserved to have its position precisely determined. Finally, the transportation problem would be easily solved since the island was on the route of the ships of the Compagnie des Indes, and there would be no need to depend upon negotiations in foreign courts.[162] The increased safety and improved facilities expected from the choice of Rodrigue were not great enough to offset the anxieties of the Académie for the health of their scientific missionary in his voyage to distant parts—or more importantly, for the possibility that illness or death might result in no observation of the transit. The assistant which they had originally assigned to Pingré for the African voyage was accordingly kept on.

Not all the French desires respecting overseas expeditions were fulfilled. The failure to send Chabert to Cyprus or to move the Spanish Court to plant an observer in rumored islands in

[161] "Du Passage de Vénus sur le Soleil; Annoncé pour l'année 1761," *Histoire*, 1757, p. 92.
[162] *ibid.*, p. 79.

the south Pacific Ocean have already been mentioned. Whether or not the lack of success was due to the lateness of the hour, the inertia of the Spanish, or the minor importance of the stations themselves is at best a conjectural matter. There was time enough to get to Cyprus, so that the first reason seems to be very weak. As for the Spanish, their interest in the transit of Venus was not awakened until 1769, and then not with a primary interest in the astronomical problem or even an eye to exploratory navigation in the vast expanse of the southern Pacific Ocean.

The ubiquitous activity of Delisle in transit matters reveals the still-born efforts of the French to get yet another expedition under way with Dutch cooperation. Ever anxious to insure the maximum possibilities for success in the observation of the phenomenon, Delisle sought to exploit his long-standing contact with Dutch astronomers by having a French observer placed at Batavia, a plantation of the Dutch East India Company. Well before June of 1760 he had written Dirk Klinkenberg, his astronomer-correspondent at The Hague, both to send him a copy of the *mappemonde* and to initiate the chain of requests designed to achieve that purpose.[163] In the meantime, while the request for permission and transportation in a Dutch bottom was pursued at The Hague, Delisle wrote to St. Florentin at Versailles with the complementary proposition in mind. Suggesting that he had already arranged things with the Dutch (which of course he had not) and stressing the importance of this particular station for geography and navigation as well as astronomy, he asked for the King's support for the enterprise. But there were also specific conditions to consider.

In Delisle's eyes the man for the job was his own assistant, Charles Messier, who had been attached to him with a £500 stipend when, shortly after his return from Russia, he had himself received an appointment as *Astronome de la Marine*. While praising his virtues and his skills, Delisle suggested in passing that Messier had long deserved an increase in salary. Delisle wished to obtain not only the King's permission and the Sec-

[163] Letter from Klinkenberg to Delisle, dated at The Hague, 6 June 1760, Dép. Marine, Vol. 115, xvi-8, No. 178b, fols. 1-3.

retary's agreement for Messier's voyage but even more their joint guarantee that his post would be held for him, and "s'il est possible de lui faire avancer deux années de ses appointemens."[164] To make it even more possible for Messier to go, Delisle indicated that he had arranged to have his work at the Observatoire de la Marine continued without interruption, even allowing for the removal of those instruments which Messier might wish to take with him.[165] But all this activity came to naught. That August day in which Chabert and La Caille gave the report drawn up for St. Florentin on the projected Pingré voyage also brought news of English intentions to observe at Bencoolen. There was no need for two expeditions to the Dutch East Indies, and the following day Delisle wrote to the Secretary of State to cancel the request for Messier's Batavia voyage. The same letter suggested that Messier accompany Pingré, whose approval of the idea Delisle had already obtained, but this plan did not materialize and Delisle's young assistant made his reputation at a later epoch and in another branch of astronomy.[166]

By the winter of 1760-61 French preparations for transit observations beyond the borders of France were thus well advanced, including those for the short trip of Cassini de Thury to Austria, there to observe at the Jesuit Observatory in Vienna, in the presence of the Archduke Joseph.[167] Preparations to observe in France, it goes without saying, were also undertaken, but these were not extraordinary in any way, and will be discussed in connection with the transit results. From June 1760 to June 1761, Delisle's private correspondence was devoted almost exclusively to the coming transit of Venus.[168] Letters on the *mappemonde*, on techniques of observation, on lenses and telescopes, as well as those simply announcing the transit or

[164] Letter to Comte St. Florentin, 16 June 1760, Obs. de Paris, B.1,1-8 (XIV-XV), 144.124.

[165] *ibid.*, fol. 1.

[166] Letter to St. Florentin, 21 August 1760, *ibid.*, 144.120.

[167] Cassini de Thury, "Observation du passage de Vénus sur le Soleil, faite à Vienne en Autriche," *Mémoires*, 1761, pp. 409-411. Father Hell observed this transit in Cassini's proximity, and the Archduke also took his turn at the telescope.

[168] Dép. Marine, Vol. 115 (XVI-8) Nos. 178-228. These letters alone, exclusive of deposits in other archives are almost wholly devoted to the transits during this span of time.

discussing it in general terms, were dispatched every few days to a world-wide audience. It seems all the more fitting, therefore, having first looked at the record of the French plans for the transit of Venus through Delisle's eyes, to bring that early phase of the transit story to a close in the same manner. But France was not alone in her intense interest in the transits; in the midst of a world war, four years old in 1761, there was room for scientific cooperation with her principal enemy, the British Crown.

BRITISH PREPARATIONS IN 1761

THE idea that the transits of Venus might serve as the natural astronomical instrument by means of which the actual scale of the universe could be laid out was due to Edmund Halley. Indeed Harlow Shapley and Helen Howarth suggest that this was one of his most important achievements.[1] But for a long time following its announcement, it seemed that British pride in that discovery would not be matched by British preparation for its full utilization.

The possibility that Mercury transits observed in the manner which Halley had prescribed for Venus might also serve to determine the astronomical unit seemed never to have evoked much enthusiasm among British astronomers. In this the authority of Halley may have, albeit unwillingly, imposed a limitation of growth upon parts of British astronomy comparable to that exercised by Newton upon British mathematics. Be that as it may, the approach of British astronomers to the transits of Mercury between 1723 and 1753 was rather straightforward and relevant only to the theory of the planet.[2] Upon this action the muse of history was to render favorable judgment for, as we have already seen, the transits of Mercury failed to yield an adequate value for the solar parallax. But Clio's favor was narrowly given, for by continuing to consider the use of Mercury in the pursuit of a definitive solar parallax, continental astronomers, particularly the French, not only developed an additional method to exploit the transits but also discovered corrections to be made in the data upon which their colleagues across the Channel intended to rely.

[1] H. Shapley and H. E. Howarth, *A Source Book in Astronomy*, p. 94.

[2] In the transit of Mercury reports recorded in *Phil. Trans.* for these years, no mention is made of Venus nor is any evidence given, from Halley's in 1723 to Short's in 1753, that anything of value for the transits of Venus could be drawn from those of Mercury.

The British were made aware of these changes through the acquisition of European memoirs, new and enlarged editions of older astronomical tables, and most rapidly through the mutual exchange of letters. With respect to the latter, Delisle's correspondence with Bradley and Bevis may again serve as an example. Some popular consideration was given to the transit of Venus in this period in the *Gentleman's Magazine*, which published Fontenelle's *Éloge* of Halley in 1747; a letter, printed in 1758, on the calculations required for the 1761 transit; and, significantly enough, a summary of Le Gentil's important paper on the Venus transits of 1761 and 1769 from the *Mémoires* of the Académie des Sciences.[3] But by and large there was either no sustained interest in the transits of Venus via those of Mercury between Halley's papers in the *Philosophical Transactions* and the advent of the Venus transit, or at best, whatever interest was expressed merely repeated what had already been said, such as a discussion of William Whiston's broadside or John Alexander's insight in 1753. Our study of the British approach to the Venus transits thus can safely begin with the preparations in 1760. And perhaps nothing can better substantiate the point just made than the fact that it was the presentation of Delisle's *mappemonde* and memoir to the Royal Society at the meeting of 5 June 1760 that precipitated British preparatory actions.[4] Delisle had dispatched the memoir from Paris on 27 April 1760, three days before the Académie des Sciences received it formally. The long delay in its arrival and presentation at the Royal Society was very likely due to the circuitous path it had to take as a result of the war between France and England.

The memoir was read to the membership and the *mappemonde* circulated. Apparently, the whole matter was fully dis-

---

[3] "The Elogy of Dr. Halley," *Gentleman's Magazine*, xvii (October 1747), 455-458 and (November 1747), 503-507; T. Fisher, Untitled letter to the editor on the transit of Venus, *ibid.*, xxviii (August 1758), 367-368; and *ibid.*, xxix (January 1759), 23-26.

[4] Journal Book of the Royal Society (hereafter cited: JBRS), xxiii, 894-899. I do not mean to suggest that the action was brought about solely by the Delisle memoir, for the June issue of *Gentleman's Magazine* (xxx, 265-269) carried a translation of Halley's dissertation on the transits of Venus. There were doubtless other stimulants.

cussed and a resolution passed to refer it to the Council of the Society for action, for by the end of the month the practical organization of the British expeditions was under way. At the meeting of 26 June 1760, with Lord Cavendish in the chair, the Council took under consideration the problem of sending "proper persons to proper places, to observe the approaching Transit of Venus . . . the Several Instruments that will be necessary for the work; and the Expenses which are likely to attend it."[5] In spite of the circumlocutions of an eighteenth-century style, and in contrast to the French records, the historical record left by the efforts of the Royal Society on behalf of the transits of Venus is a model of clarity. With this in mind, together with the historian's assumption that organized human activity for common ends will, in the same epoch and within the same historical tradition, take roughly similar forms,[6] it should be possible to arrive at a more complete understanding of the French and other national preparatory techniques in the light of the more distinct record of British activities, even where the well of direct documentary evidence for the others runs dry.

The consideration given to the transit problem that same June day was brief and to the point: the Council simply decided that it was both proper and expedient for the Royal Society to direct the observations. And with this decision behind it, the Council immediately proceeded to the remaining items of business relevant to that actual performance. Obviously, the choice of overseas stations was the first order of business, with two of the most-favored points of observation to be selected from the original list which Halley had published in the 1716 paper in the *Philosophical Transactions*[7] and the subsequent changes which Delisle had argued for in the memoir joined to

[5] Council Minutes of the Royal Society (hereafter cited: CMRS), IV, 228-229.
[6] For example, the raising of funds for large-scale scientific enterprises in the eighteenth century, invariably took the form of petitions to the Crown or the lesser ranks of royalty. In this respect, it will be interesting to compare the specific prayers for aid to their sovereigns by the Royal Society of London and the Académie des Sciences.
[7] E. Halley, "A New Method of Determining the Parallax of the Sun, or his Distance from the Earth," *Phil. Trans. Abgd.*, VI, 245. The places cited by Halley were: Trontheim (Norway), the gulf of the Ganges, the Kingdom of Pegu, Fort St. George (Madras, India), Hudson's Bay, Bencoolen, Pondichery, and Batavia.

the *mappemonde*. On this basis, first choice went to the island of St. Helena in the South Atlantic Ocean, and second to Bencoolen on the East Indian island of Sumatra, with Batavia listed as an alternate for the latter selection "if it were not attended with uncertainty."[8]

Dr. James Bradley, successor to Halley at Greenwich since 1742, was present at the meeting and was called upon to list the instrumental needs which the proposed expeditions would require in order to perform their tasks adequately. This opinion of the Astronomer Royal in 1760 is especially informative, for it represents as authoritative a voice as can be summoned from the past to speak on the kind of portable instruments required by an astronomical expedition. To some extent his enumeration can be taken as the minimum standard for all expeditions. The list included a "Reflecting Telescope of Two Foot with Dolland's Micrometer, Mr. Dolland's Refracting Telescope of Ten Feet; a Quadrant of the radius of Eight Inches; and a Clock or time piece."[9] Instruments were expensive and not too easy to come by at this time. In spite of the already distinguished work by men like John Dolland, John Bird, and Jesse Ramsden in the design and manufacture of scientific instruments, especially those devoted to navigation and astronomy, the techniques of large-scale production had yet to take hold in England.[10] It would not, perhaps, be premature at this point to suggest that the interest in multiple observations occasioned by the transits of Venus (and by other astronomical problems on a similar, albeit lesser scale, such as the various measurements of a degree of the meridian) helped to introduce such techniques into English ateliers well before the century had run its course. One should not read into that statement, however, an argument for the unique importance of astronomical problems in producing a miniature industrial revolution in the manufacture of scientific instruments. Of the great scientific problems of the age,

8 CMRS, IV, 229.

9 *ibid.*, pp. 229-230.

10 M. Daumas, *Les Instruments scientifiques aux XVIIe et XVIIIe siècle,* p. 126. ". . . [V]ers 1780, les lunettes de Dolland, fabriquées en série dans un atelier où était établie la division du travail en plusieurs opérations exécutées toujours par les mêmes ouvriers, se vendaient à profusion."

it was certainly those connected with astronomy which made the greatest demands upon the skill and productive technique of the instrument maker in the high noon of the eighteenth century. The determination of the solar parallax, the theory of the shape of the earth and, indirectly, the determination of longitudes at sea, and the geographical exploration of the globe support that contention. On the other hand, the discovery of electricity and the rising tempo in the development of chemistry brought new kinds of instruments into being and accelerated their circulation. Thus the burgeoning industry in scientific instruments in the last half of the eighteenth century was itself undergoing the kind of industrial revolution in specialization, division of labor, and narrow concentration that was then taking place in the mining industry and that had already occurred in the manufacture of textiles in the first half of the century.[11]

But the kind of mass production which would eventually bring about a significant reduction in the price of instruments lay just beyond the immediate horizon of those faced with the problem of obtaining them in 1760. A good achromatic telescope by Dolland, of the kind which had first been presented to the Royal Society only two years earlier by Short,[12] is estimated to have cost about £1400.[13] The meaning of that price will perhaps be more fully understood when contrasted with Delisle's annual salary, as *Astronome de la Marine*, of £3000, or Messier's, as his assistant, of £500. It is not at all surprising to discover, therefore, that the initial intent of the Royal Society was actually to hire the instruments required, and to that end the Council requested Bradley "to inform himself against the next Council, upon . . . [the] Terms."[14] The last item of business undertaken at this extremely efficient meeting of 26

---

[11] See P. Mantoux, *The Industrial Revolution in the Eighteenth Century* (London: J. Cape, 1948), pp. 193-224 and 277-312.

[12] *Phil. Trans.*, 1758, p. 733. The five-foot achromatic refractor which was presented that day (8 June 1758), gave as sharp and bright an image as earlier fifteen-foot refractors. A note of secrecy was also maintained, for no details capable of being copied were read into the record.

[13] Daumas, *op.cit.*, p. 191. These of course are French pounds and should not be confused with English pounds to be used in later British accounts.

[14] CMRS, IV, 229.

June 1760 centered on the personnel to be selected and the transportation to their destinations. Through one of the members present an inquiry was addressed to the East India Company, involving passage to and from the stations chosen, the accommodations to be expected there, and other travel miscellanea, such as ship schedules, costs, and the like. It is in keeping with the British tradition that, in contrast with the expeditions of other nationals, the first request for assistance should be addressed not to the Crown but to John Company. Apparently the question of who was to staff the expeditions could not be decided upon immediately, and the decision was deferred until another time, the membership being invited in the meantime to consider qualified individuals known to them and to obtain the terms under which they might agree to go.[15]

Within the week following this crucial series of decisions and before the next meeting of the Council, two reports pertinent to the transits were received by the Royal Society. They reveal the care which went into the preparation of an expedition, and a measure of the interest which the transit problem had evoked in nonscientific circles. Furthermore, they are equally significant in highlighting the reservoir of talent and experience for this sort of activity which Britain had accumulated by the spring of 1760. In this respect a word on Britain at mid-century will not be amiss.

In October of 1760, when George III succeeded his grandfather, inheriting along with his crown the political brilliance of William Pitt, English power stood at its peak. The Seven Years' War with France, that first of global wars, was only about half over, but British bells, as Horace Walpole had remarked, were "worn threadbare with ringing for victories."[16] The respect for Britain among the nations of the world was perhaps at its highest in history. No virulent anti-English tradition within the Empire had yet arisen to troublesome proportions: the Irish lay in rural quiet, momentarily forgotten; the American colonies were still close to the expanding bosom of the realm and de-

[15] *ibid.*, p. 230.
[16] Quoted by Lord Elton, *Imperial Commonwealth* (New York: Reynal & Hitchcock, 1946), p. 164.

voted to the Great Commoner. Indeed, the Americans had played no small role in that expansion into the North American hinterland, and now looked to the Crown for special privileges in the area where they had fought. Even France, victim of the ubiquitous British navy and the East India Company's long secular arm in the Orient, admired England as much as she feared her, for much of the nation was out of step with the despotic institutions and the inept leadership which had brought her to such a sorry pass.

Yet in the midst of war for commerce and empire there was time in Britain for scientific enterprise and international cooperation. The first and more important of the two reports referred to earlier gives substance to that statement, and does so strikingly, for the issues of conflict were not only national in character but involved those of great personal ambition as well. When the Royal Society's spokesman approached Laurence Sullivan, Chairman of the Court of Directors of the East India Company, for assistance in placing observers on St. Helena and in the East Indies, he did so at a key moment in the lives of the Company and its Chairman. The crucial and bitter election of 1758, the first to be greatly contested in the East India Company's long history, had just given Sullivan the chairmanship of the Court of Directors.[17] Shortly thereafter (1759), recognizing that Sullivan's move was a bid for absolute power, Clive of India took umbrage at this attempt to circumscribe his own ambitions within the Company.[18] And one year later, in July of 1760, when Clive returned to England, the battle between the two was joined.[19]

In the face of this challenge to his recently-acquired power, Sullivan calmly met with the Royal Society's representative and other directors of the Company at the East India House on 2 July 1760 to discuss the question of assistance to the Society's observers.[20] The details of this conversation were reported at the

[17] L. S. Sutherland, *The East-India Company in Eighteenth-Century Politics* (Oxford: The Clarendon Press, 1952), p. 71.

[18] *ibid.*, p. 76.   [19] *ibid.*, p. 84.

[20] CMRS, IV, 237-243. "Memorandum . . . of a Conversation at the East India House with Mr. Sullivan (Chairman) Mr Rous (Deputy Chairman) Mr Boyd (late Deputy Chairman) Mr Christopher Barrow and Mr George Amyand (Di-

Council meeting of the Royal Society which took place the next day. Essentially, the report revealed an attitude of full cooperation on the Company's part which went beyond the point of simple accommodation. They promised to do all within their power to facilitate the affair at any of their settlements appropriately located, and indeed, had already "at the Sollicitation of some members of the University of Cambridge [sent] instructions to all their Presidencies relative thereto."[21] With the season well advanced, the question of getting the expeditions off became paramount. The voyage to St. Helena would take only three months, and scheduled sailings to the island were such as to make possible the departure and return of the expedition within a year. Or in any case, it could be regulated by the length of the stay which the observers chose to make at the island.[22] But so far as the Company was concerned, there was absolutely no chance at that date of getting to Bencoolen in time to observe the transit on 6 June 1761, and to make the companion observations necessary to ensure its success. A suggestion was made that Dutch ships might make it to Batavia in time, but no action was taken upon it. An additional observational station, Madras in India, was discussed, but not with the intention of sending observers there, "because there . . . [was] a probability, tho not a certainty of a conveyance in due time by the Co.'s Ships."[23] According to Sullivan, however, there was a man at Madras named Call who was capable of making the observations, providing the Cambridge instructions got to him in time. It was also likely that he, or someone accessible to him, possessed the one set of instruments in all of India required for the job.[24] Finally, the Company advised the Society that because of the war with France as well as the losses of Indiamen to privateers, the hazards of travel had multiplied greatly and that, therefore, if two or more persons were sent to the same place, they should "go in *different Ships* . . . to

---

rectors) concerning the Times and places of observing the Transits of Venus on the Sixth of June 1761. and the assistance and encouragement to be expected from the Directors of the East India Company."
[21] CMRS, IV, 232.   [22] *ibid.*, p. 238.
[23] *ibid.*, p. 231.   [24] *ibid.*, p. 232.

[avoid] the Risque of both being Embarked on the *Same* Bottom."[25]

The instructions for observing the transit of Venus which unknown hands at Cambridge University had put together for distribution by the East India Company, well in advance, it must be admitted, of the Royal Society's petition for assistance,[26] are a model of clarity and simplicity. These characteristics and the pattern of its distribution give it a high measure of importance. Addressed as it was to amateurs and circulated to the far corners of the world, it is partially responsible, with companion pieces to be found in publications like the British *Nautical Almanac* and the French *Connoissance des temps*,[27] for the abundance of nonprofessional observations of both transits everywhere. Some of these amateur observations were of a high enough caliber to find their way into contemporary scientific journals, but many of them remain recorded only in the pages of diaries and personal letters such as those of Ezra Stiles and Manassah Cutler in America in 1769—mute evidence of the breadth of active interest awakened by the transit problem and the quality of instructions issued to handle it.

The second report which the Royal Society received in the interval between the Council meetings of 26 June and 3 July 1760 concerned the physical conditions to be expected at Bencoolen. It came from the experienced hands of Governor Lennox, "who [had] resided there 16. or 17. years."[28] Mostly an analysis of the weather conditions on the island, it emphasized the uncertainty of the rainy season and the presence of steady

---

[25] *ibid.*, p. 239.

[26] These instructions had already been dispatched by May 1760 to the Company's Presidency at Madras on the *Chatham* man-of-war. Royal Society, Miscellaneous Manuscripts (hereafter cited: Roy. Soc., Misc. MSS), x, 106, fol. 1.

[27] See for example, the *Connoissance des temps pour l'année 1761* (Paris: Imprimerie Royale, 1759), Vol. 2, pp. 145-156, for Lalande's statement in the same vein, and Maskelyne's for 1769, originally published in the *Nautical Almanac and Astronomical Ephemeris for the Year 1769*, but separately printed as *Instructions Relative to the Observation of the Ensuing Transit of the Planet Venus over the Sun's Disk, on the 3d of June 1769* (London: Richardson & Clark, 1768).

[28] "Result of Inquiries concerning Bencoulen, Batavia, & St. Helena, relative to the taking Observations in those Places of the Transit of Venus on 6th June 1760," Roy. Soc., Misc. MSS, x, 104, fol. 1. In spite of the title the report is limited to Bencoolen, with only passing reference to the other places.

winds for months at a time. But it also added a rather revealing comment on the nature of British colonization in the Orient, one that is suggestive of the Australian pattern to come. "The place is now *wholesome*," Lennox wrote, "and inhabited by English of *better Character & Fashion*, than formerly; when it was unwholesome and peopled by the Refuse of our Countrymen in those parts."[29]

The information obtained from both the East India Company and Governor Lennox was thus presented to the Council on 3 July 1760. At the same meeting Bradley reported the impossibility of hiring instruments for the occasion, and an estimate for the cost of instruments required was accordingly submitted by Nevil Maskelyne. Although he was to become Astronomer Royal in only five years, this was Nevil Maskelyne's first opportunity to take part in a large-scale astronomical enterprise; and it was largely as a result of Bradley's influence that he was able to enter into the deliberations on the transit problem.[30] Apparently Maskelyne had earlier suggested that it would be advantageous to make additional observations at St. Helena for the purpose of discovering the annual parallax of Sirius, a discovery which would, among other things, give actual proof of the validity of the Copernican system.[31] Consequently, when he presented the list of equipment needed, a special instrument for that task was included. The total cost as Maskelyne figured it, came to £285, and it would be worthwhile to itemize the components of that figure. Three 2-foot reflecting telescopes for the transit of Venus observations were ordered, two for St. Helena and one for the corresponding observations at the Royal Observatory at a cost of £35 each. Two of these telescopes were to be fitted with Dolland's micrometers to convert them into instruments of high precision; cost £30. A pendulum clock "adjusted so as to obviate the variable effects of heat and cold" at St. Helena was acquired for £20. This same clock, having been associated with many scientific expeditions since the transits of

---

[29] *ibid.*, fol. 2.

[30] M. A. Clerke, "Nevil Maskelyne," *Dictionary of National Biography* (hereafter cited: DNB), XII, 1299-1301.

[31] See N. Maskelyne, "A Proposal for discovering the Annual Parallax of Sirius," *Phil. Trans. Abgd.*, XI, 501.

Venus, is still keeping good time in a corner of the Royal Society's London office.[32] Finally an 18-inch quadrant was acquired (at £30) for observations designed to regulate the timepiece, and a 10-foot astronomical sector, constructed by Dolland with a compound objective, to be used for the Sirius project; price, £100, the most expensive instrument of them all.[33]

To this sum the Council then added the other expenses of the St. Helena expedition: transportation, subsistence, and remuneration for the personnel involved. Passage to and from the island came to £120 for the two men. The principal observer, who was also to undertake the observation of Sirius, was allowed £100 for ten months' expenses, and his assistant, who was to remain at St. Helena for only the main project, that is, for about four months, was allowed £30. A gratuity of £100 would be given the principal observer and £50 to his assistant. The total cost of the St. Helena expedition was thus estimated at £685, "which Sum being judged to be greater than was Convenient for the Society to Expend a Memorial was drawn up, to be presented to the Lords of the Treasury."[34]

One of the themes most stressed in this memorial was the competitive and nationalistic aspect of the undertaking. Perhaps this was a reflection of the war in progress, or perhaps it was merely the correct psychological approach to the Lords of the Treasury. Whatever the reason, the fact that the French and other European courts were sending astronomical expeditions to various parts of the world was felt to be of immediate concern to the British Empire and was viewed with some apprehension about national prestige. For here was an enterprise wholly initiated by an Englishman in the previous century, and even more, "never observed but once before Since the World began, and then only by another Englishman, the Ingenious Mr.

---

[32] See T. D. Cope, "A Clock Sent Thither by the Royal Society," *Proceed. Am. Phil. Soc.*, Vol. 94, No. 3 (1950), 260-268; H. A. Lloyd, "Description of a Clock by John Shelton, Owned by the Royal Society and used by Maskelyne on his Visit to St. Helena in 1761, and Probably by Mason and Dixon in Pennsylvania," *ibid.*, pp. 268-271; and H. A. Lloyd "A Link with Captain Cook and H. M. S. *Endeavour*," *Endeavour*, x, 40 (October 1951), 200-205; also the *New York Times*, 23 March 1952.

[33] CMRS, IV, 242.

[34] *ibid.*, p. 233.

Horrox."[35] Possibly because they expected some trimming to take place, or perhaps because they were simply rounding out figures to include unforseen emergencies, by the time the Society had formally drawn up its petition, the earlier £685 estimate had grown to £800 in the actual memorial. Furthermore, this covered only the St. Helena expedition, and the memorialists indicated that a similar mission to Bencoolen, which was very much to be desired, would bring the total "to near double that Sum,"[36] which in its original size had been well beyond the Society's means.

The petition was drawn up and appropriately signed and sealed by the President and Council the next day (4 July 1760), but the chances of success were not left to rest upon the strength of the memorial's argument alone. Lord Macclesfield, the Society's president, undertook to strengthen the request by an appeal to the Duke of Newcastle. He could not have done better, for Thomas Pelham-Hollis, first Duke of Newcastle, whom Trevelyan called "the greatest borough-monger England ever produced,"[37] was the most skillful political manager of the age. Even the great Pitt, for all his popularity had had to form an alliance with Newcastle to stay in office. For it was Newcastle who insisted that Pitt be made Secretary of State in 1757 as the price of his own return to the government after George II's demonstrated impotency as a Cabinet maker.[38] The separate appeal to Newcastle repeated the arguments in the name of patriotism and honor to be found in the memorial, but with a heightened emphasis in language and color that is worth quoting.

Apparently Macclesfield's original intention had been to argue the case orally for he had visited Newcastle to present him with a copy of the memorial, but finding him unavailable, was forced to record his points in a letter dated 5 July 1760, only one day

[35] "To the Right Honourable the Lords Commissioners of his Majesty's Treasury. The Memorial of the President, Council, and Fellows of the Royal Society of London for Improving Natural Knowledge," ibid., p. 234.

[36] ibid., p. 235.

[37] G. M. Trevelyan, A Shortened History of England (New York: Longmans Green & Co., 1944), p. 374.

[38] See D. L. Keir, The Constitutional History of Modern Britain, 1485-1937 (London & New York: D. Van Nostrand, Inc., 1948), pp. 319-320, 333, and 336.

following the signing of the original petition. This accident and the fortunate preservation of the letter have therefore made it possible for us to learn the kind of argument on behalf of this scientific enterprise that would carry weight with high government officials in the middle of the eighteenth century.

My Lord Duke:

I did myself the honour this morning of calling at Your Grace's door, with an intent to lay before you the enclosed Copy of a Memorial from the Royal Society, which, I believe will be presented to the Lords Commissioners of the Treasury at their next meeting. . . .

The Memorial itself plainly shews, that the Motives on which it is founded are the Improvement of Astronomy and the Honour of this Nation; which seems to be more particularly concerned in the exact observation of this rare phaenomenon, that was never observed but by one Englishman . . . and . . . pointed out and illustrated by Dr. Halley another Englishman. And it might afford too just ground to Foreigners for reproaching this Nation in general (not inferior to any other in every branch of Learning and more especially in Astronome); if, while the French King is sending observers . . . not only to Pondicherie and the Cape of Good Hope, but also to the Northern Parts of Siberia; and the Court of Russia are doing the same to the most Eastern Confines of the Greater Tartary; not to mention the several Observers who are going to various Places, on the same errand from different parts of Europe; England should neglect to send Observers to such places . . . subject to the Crown of Great Britain.

This is by foreign Countries in general expected from us; Because the use that may be derived from this Phaenomenon, will be proportionate to the numbers of distant places where . . . observations . . . shall be made of it; And the Royal Society, being desirous of satisfying the universal Expectations of the World in this respect have thought it incumbent upon them . . . to request your effectual intercession with His Majesty . . . to enable them . . . to accomplish this their desire . . . which . . . would be attended with expense disproportionate to the narrow Circumstances of the Society.

But were the Royal Society in a much more affluent State; it would surely tend more to the honour of his Majesty and of the Nation in general, that an Expense of this sort, designed to promote Science and to answer the general Expectation of the World, should not be born by any particular Set of Private Persons.[39]

Five days later, at the Council meeting of 10 July 1760, at which time Macclesfield read the gist of this letter to those present, it

[39] Roy. Soc. Misc. MSS, x, 109, fols. 1-3.

was learned that some additional instruments would be loaned to the Society by Ellicott and Bradley for the use of the expeditions. At the same time Bradley suggested that his assistant serve as a second party for the St. Helena expedition.[40]

When the Council met again a few days later, Macclesfield's wisdom in writing the Duke of Newcastle became apparent. For one thing, the original memorial had been improperly addressed, and Newcastle, recognizing the need for a rapid decision on financial support if adequate preparations were to be made, had simply not shown it to the Lords of the Treasury but carried it directly to the King. Whether Newcastle's decision to hurry the matter along was due to the patriotic polemic in Macclesfield's appeal or merely to a sympathetic desire to help the Royal Society satisfy the "universal Expectations of the World" is unimportant. What counted were the speed and the success of his actions, for by 14 July 1760 the Council happily learned that a warrant for £800 to cover the expenses of the St. Helena expedition was immediately available. A second warrant for the same amount would be issued at a later date if it could be shown that the expedition bound for Bencoolen would be likely to arrive there in time.[41]

The response to this good news was immediate. Resolutions were passed to employ Nevil Maskelyne as principal and Charles Mason,[42] Bradley's assistant, as second in the proposed St. Helena observations. The generosity of the Crown in supplying the full amount also made it possible to increase the stipend of both observers, Maskelyne's from £100 to £150, and Mason's from £50 to £100. Another resolution was passed giving Maskelyne the power to order the instruments listed earlier on 3 July 1760, some of them from the workshop of Jonathan Sisson.[43] In response to the Crown's attitude toward the observations at

---

[40] CMRS, IV, 247.   [41] ibid., p. 249.

[42] Biographical material on Mason, beyond the meager account in the DNB, can be obtained from H. W. Robinson's "A Note on Charles Mason's Ancestry and His Family," Proceed. Am. Phil. Soc., Vol. 93, No. 2 (1949), 134-136, and the innumerable articles of T. D. Cope on Mason and Dixon in the same publication and others such as The Scientific Monthly and Pennsylvania History.

[43] Jonathan Sisson (1690-1760) was the principal collaborator of George Graham, in whose workshop he was long employed. He was one of the finest mechanics of his day, and his instruments were noted for the finish and elegance of their construction. See Daumas, op.cit., pp. 304-306.

Bencoolen, the Council drew up a rough estimate of expenses for this second expedition. Exclusive of payments to the men involved, it came to about £370, covering the cost of instruments (two reflecting telescopes, one with a micrometer), transportation, and subsistence.[44]

Through July and August of 1760, the Council of the Royal Society continued to be preoccupied with the transit of Venus expeditions, and in that interval several important actions were agreed upon. On 21 July 1760 a petition was drawn up and addressed to the Admiralty on behalf of the voyage to Bencoolen. Although it had been the Society's original intent to make full use of the East India Company for the expedition to Sumatra as well as the one to St. Helena, the anxiety of the Society's officials over the little time remaining for preparation and dispatch of the observers, together with the knowledge that the scheduled sailings of East Indiamen for Sumatra were not all that could be desired for ultimate success, led the Council to petition the Navy for the appointment of a man-of-war for the task. Couched in the usual flattering terms, the request was strengthened by the King's earlier financial support of the whole enterprise as well as the bid to the Lords Commissioners of the Admiralty that they would "want no further Sollicitation to give this new Instance of . . . [their] Zeal for the promotion of a Science so intimately connected with the Art of Navigation as well as for the Honour of the Nation."[45] And by 5 August 1760, the Council was informed that a ship would be provided for this purpose.[46] In the same period, several different estimates of expenses for the St. Helena trip were submitted by Maskelyne and Mason, but however they varied, up or down, a constant increment in their expenses was the estimated cost of liquor.[47] It is difficult to say whether this was a result of increasing knowledge about the dangers of the long voyage in wartime and the loneliness of St. Helena, or simply a desire to have the best of

[44] CMRS, IV, 250.

[45] *ibid.*, p. 254.

[46] Letter to Dr. Birch from Mr. Cleveland, Secretary to the Lords Commissioners of the Admiralty, CMRS, IV, 256.

[47] Roy. Soc., Misc. MSS, X, 111, and CMRS, IV, 251, 257-258. In keeping with his rank in the expedition, Mason's liquor estimate was proportionately lower.

the two worlds implied in Voltaire's quip that "en ce bas monde il vaut mieux être gastronome qu'astronome."

Although many volunteers had written to the Council offering to make the voyage to Bencoolen, it was not until the meetings of 11 and 25 September 1760 that the personnel issue was discussed. Once again the Astronomer Royal spoke in the name of his assistant, for Bradley suggested that Mason would be willing to undertake the Bencoolen expedition as principal observer. Obviously this had been debated earlier, for the issue was immediately resolved with a payment of £200, exclusive of the necessary expenses, voted to Mason. A motion was also passed that a proposal be made to Mr. Jeremiah Dixon, surveyor and amateur astronomer of Cockfield in Durham, to accompany Mason at the same fee of £200 but in a secondary or subordinate position. Before the month was over Dixon had favorably replied to the Society's offer and he was accordingly employed.[48] It was thus the occasion of the transit of Venus in 1761 that brought into being the astronomical team of Mason and Dixon, whose names have become household words in America for their labors in Maryland and Pennsylvania in 1763 and the line which they surveyed in that region.[49]

In spite of Mason's promotion to the leadership of the second expedition, the selection of instruments was again left to Maskelyne, and this time he ordered the telescopes from James Short.[50] Maskelyne also received a new assistant, Robert Wad-

[48] CMRS, IV, 259-261.

[49] There is no mention of Dixon in the DNB, but H. W. Robinson has written "Jeremiah Dixon (1733-1779)—A Biographical Note," Proceed. Am. Phil. Soc., Vol. 94, No. 3 (1950), 272-274. Other relevant material is to be found in H. P. Hollis, "Jeremiah Dixon and his Brother," Journal of the British Astronomical Association, 44 (1934), 294-299; T. D. Cope, "Mason and Dixon— English Men of Science," Delaware Notes, 22nd Series (1949), 13-32; T. D. Cope and H. W. Robinson, "Charles Mason, Jeremiah Dixon and the Royal Society," Notes and Records of the Royal Society of London, IX (October 1951), 55-78, and numerous other articles by the latter two authors to be found in the final bibliography.

[50] James Short (1710-1768) was one of the most prolific producers of telescopes among the English instrument makers. During the second half of the eighteenth century, most European observatories had a Short telescope of one kind or another. He was a member of the Royal Society, had studied mathematics with Maclaurin, and had published articles on Saturn and its rings. Daumas, op.cit., pp. 226-227, 303-304. The clock used by Mason and Dixon was built by John Ellicott, F.R.S. It was to have a history almost as involved as John Shel-

dington, to replace Mason on the St. Helena expedition. With the personnel for both expeditions complete and the equipment on order by October, the Society turned again to the East India Company to obtain more specific information about Bencoolen and a definite commitment from the Company. In a letter to Laurence Sullivan written on behalf of the Society, Charles Morton made several interesting requests. What assistance exactly could the observers expect from the Company's agent at Bencoolen? What timber and workmen were to be found there "for Erecting a Slight Building for the Play of Instruments and . . . observations"? Since the man-of-war which was to bring Mason and Dixon to the island was afterwards "to be Employed . . . in the Public Service," how soon would one of the Company's bottoms touch at Bencoolen on a homeward voyage? Finally, it was the desire of the Council to meet with a representative of the Company to whom Bencoolen was familiar, if such a person were available in London.[51]

The Company's reply was not only immediate but more than satisfactory. One of their men, a Captain Wilson "lately at Sumatra," was able to inform them of men and materials available for constructing the observatory, and to add that the weather was usually excellent during the time in question. The observers would be accommodated by the Governor and Council of Fort Marlborough "with Diet and Apartments" on the west coast of Sumatra, and, when the time came, with passage home —all at the Company's expense. The only drawback to the latter was that they might have to return via the Company's plantations in China or Bombay, for no ship would go to England directly from Sumatra.[52] In spite of the latter condition Mason and Dixon still agreed to go, and on 23 October 1760 they signed the Articles of Agreement legally binding them to the mission. About the same time they signed a receipt for the instruments

---

ton's clock used by Maskelyne. See T. D. Cope and H. W. Robinson, "Charles Mason, Jeremiah Dixon and the Royal Society," *op.cit.*, p. 57, and CMRS, IV, 260, 262, 270.

[51] CMRS, IV, 265.

[52] Roy. Soc., Misc. MSS, X, 114. Letter from Robert James, Secretary of the East India Company, dated East India House, 16 October 1760, to Charles Morton.

which is a valuable catalog of those which they actually took along.

Throughout November and well into December last-minute preparations for the expeditions were hurried along. Most pressing of all was the Bencoolen voyage, the longer of the two and the one most likely to run into trouble as a result of the war with France. The problem of getting the instruments completed in time seems to have also increased the anxieties of all concerned, especially in the matter of John Ellicott's clock; but by 13 November Ellicott reported to the Council that it was ready to go at a day's notice, and Mason also informed them that all the other instruments were packed and ready to go, being temporarily "lodged in a hired Room at the White-Hart in the Burrough, ready for the Portsmouth carrier." In addition he informed them that he would be at Portsmouth on the 20th to receive the instruments and to board H. M. S. *Seahorse* which the Admiralty had assigned to the voyage.[53] Several lesser details of the voyage were arranged in the same period. These included bills of credit with the East India Company for emergency use by each of the observers, arrangements with the Collector of Customs at Portsmouth for the transfer of the instruments to the ship, and finally, an agreement with Captain Smith of the *Seahorse* as to the victualing of Mason and Dixon with the ship's officers, down to the last detail of assuring them their turn at the Captain's table "as the Custom was, among the principal officers."[54]

But while the Council was busily engaged in the preparation and dispatch of the expeditions, the Society as a whole at its meetings was not idle. It received a paper on the approaching transit of Venus[55] from Father Roger Boscovich (1711-1781), one of the most interesting of the Jesuit astronomers at work in Italy. He had conducted the measurement of an arc of the meridian in the Papal States, and encouraged others to do the same elsewhere. And only the year before he had published at Vienna his *Philosophiae Naturalis Theoria* (1759), which until

---

[53] CMRS, IV, 274-275.
[54] *ibid.*, pp. 277-281.
[55] "On the next approaching Transit of Venus," *Phil. Trans. Abgd.*, XI, 500.

recently has been unjustifiably neglected.[56] On 20 November the membership listened to a discourse on the transits by James Ferguson, the self-taught Scots astronomer and "peasant philosopher," who had independently, it seems, hit upon Delisle's method for using the transits of Venus.[57] Ferguson's discourse was accompanied by four projections of the transit, designed to illustrate his forthcoming publication on the subject.[58]

On 8 December 1760, when James Short and Benjamin Franklin took the oath as members of the newly elected Council, Mason was aboard ship at Portsmouth, writing the Society that every thing was ready "and waited only for the wind," and, as if to assuage potential anxieties, he mentioned that the Captain doubted that Bencoolen had been captured by the French.[59] This early confidence on the part of the Captain and his astronomers was not, however, matched by the upper echelons of his Majesty's naval arm; and before the ship upped anchor, the Admiralty wrote the Society to request an alternate site in the East Indies, "in case the Sea-Horse when she arrives there, should find Bencoolen in the hands of the French." Batavia was named as the next eligible place and a letter to that effect was written to the Admiralty together with the request that "all convenient Expedition be made," because time was growing quite short.[60]

With the first expedition thus virtually under way the next order of business was accelerating the departure of Maskelyne and Waddington. Once again the instrument makers were asked

[56] See R. J. Boscovich, A Theory of Natural Philosophy, with A Short Life of Boscovich (Chicago: Open Court Publishing Company, 1922). It was his influence which led to the measurements of the meridian by Liesganig in Austria and Beccaria in Piedmont. Phil. Trans. Abgd., XI, 500, and R. J. Boscovich and C. Maire, Voyage astronomique et géographique, dans l'état de l'eglise, entrepris par l'ordre et sous les auspices du Pape Benoit XIV, pour mesurer deux degrés du méridien, & corriger la carte de l'état ecclésiastique. . . . (Paris: N. M. Tilliard, 1770).

[57] See Ferguson, Astronomy Explained Upon Sir Isaac Newton's Principles. . . . Also, E. Henderson, Life of James Ferguson, F.R.S., in a Brief Autobiographical Account, and Further Extended Memoir (2d edn.; Edinburgh: A. Fullerton & Co., 1870), and Anonymous, "James Ferguson, the Astronomer," Blackwood's Edinburgh Magazine, Vol. 134, No. 814 (August 1883), 244-263.

[58] JBRS, XXIII, 961. The pamphlet was called A Plain Method of Determining the Parallax of Venus by her Transit over the Sun (London: 1761).

[59] Roy. Soc., Misc. MSS, X, 124.

[60] CMRS, IV, 291-292.

to hurry their labors along, and bills of credit were established for the observers with the East India Company's Governor at St. Helena, not to exceed £190 for Maskelyne and £30 for Waddington. Instruments were finally delivered into their care and, for all practical purposes, the St. Helena expedition was ready to depart. But it was almost a month before they actually sailed, aboard the East Indiaman *Prince Henry* on 17 January 1761.[61] Between that date and 23 December 1760, when they are last mentioned in the records of the Royal Society, we must assume that they were occupied with personal affairs, last-minute preparations, and the actual move to board ship, very likely in the manner already described for the Mason and Dixon departure.

The apparent contradiction of having an international, cooperative scientific enterprise under way in the midst of war comes into high relief as a result of Mason and Dixon's brief naval encounter in the roiled waters of the English Channel. Barely thirty-four leagues out, the *Seahorse* was engaged by the French thirty-four gun frigate *le Grand*. A violent battle of about an hour's duration, in which Captain Smith seems to have given a very good account of himself, resulted in a mutual withdrawal. Nevertheless, the *Seahorse* returned to Plymouth harbor with eleven dead, thirty-seven wounded (many of them mortally), and some minor damage to the astronomical equipment.[62] Mason's letter describing the event also reported damage to the ship and the need for extensive repairs. The delay thus occasioned, he thought, would make it impossible for them to get to Bencoolen in time, and he therefore asked for new instructions. When this letter of 12 January 1761 was reported to the Council by the Society's secretary, Dr. Charles Morton, on 21 January, it was accompanied by assurance that everything possible would be done to prevent such an incident again. On 16 January, probably as soon as he had received Mason's letter,

[61] *Phil. Trans.*, LII, 26 and 573. Also T. D. Cope, "The First Scientific Expedition of Charles Mason and Jeremiah Dixon," *Pennsylvania History*, XII, 1 (January 1945), 5.
[62] Roy. Soc., Misc. MSS, X, 128. Letter from Mason to Morton, dated Plymouth, 12 January 1761. See also R. H. Heindel, "An Early Episode in the Career of Mason and Dixon," *Pennsylvania History*, VI, 1 (June 1939), 20-24 for a full discussion of this incident alone.

Morton had consulted the Admiralty and learned that orders
had already gone out to refit the *Seahorse* at Plymouth and have
her proceed to her original destination. To assure her of a safe
departure, the Admiralty this time arranged for a seventy-gun
ship-of-the-line to escort her out of the Channel.[63] Incidentally,
the French astronomers were informed of this incident by a
letter from Morton to Delisle.[64]

Whether it was due to the immediate incident, the rigors of
the sea freshly experienced, or their scientific anxieties about the
delay, the long voyage to Bencoolen no longer seemed as inviting or as important to Mason and Dixon as it had earlier.
Twice on 25 January and once again two days later, they wrote
to the Council of their altered intention not to go to Bencoolen.
"We will not proceed thither," wrote Mason in the first of these
letters, "let the Consequence be what it will."[65] Instead he proposed that they go to Scanderoon in the eastern part of the
Black Sea, where they might at least see the first internal contact of Venus with the sun, an observation which would probably be of use in conjunction with those expected to be made
by the French in Siberia. "Perhaps," he continued, "the Council of the Society may take it strange of me but I see no reason
why I should go upon Impossibilities. and then perhaps at my
return they will suppose the failure was owing to me, if chance
should place us there a day or two before the time. between
the Uneasiness I have given my self, to see the Uncommon
Misfortune that have attended our designs, and the sea sickness
besides; have affected me in an Unusual Manner."[66] Between
this letter and the dispatch of the second on the same day,
the intrepid observers received two from the Society instructing
them to do everything in their power "to answer the Intention"
of the expedition. Apparently these letters had also suggested

---

[63] CMRS, IV, 294.

[64] Dép. Marine, Delisle MSS, 115 (XVI-7), No. 25b. This is a later letter
(19 May 1761), but Delisle wrote on the margin: "Le vaisseau sur lequel les
premieres astronomes Anglois destines a faire les observations a Bencoolen, avoit
été si maltraité par un armateur François qu'il étoit obligé de revenir en Angleterre comme Mons. Morton m'a informe il y a longtemps."

[65] Roy. Soc., Misc. MSS, X, 129. Letter from Mason to Morton, dated Plymouth, 25 January 1761.

[66] Roy. Soc., Misc. MSS, X, 129, fol. 1.

that the voyage itself be rerouted via the Cape of Good Hope, while retaining the original goal; for in Mason's second letter, now supported by Dixon's signature, the argument was advanced that it was absolutely *"impossible . . . to reach any port by way of the Cape proper for making the Observations that will have East Longitude sufficient to be of any use. . . ."*[67] (The italicizing is Mason and Dixon's!) Once again the virtues of Scanderoon were stressed, and the two observers informed the Society that they would be pleased to go there, but would "not proceed from this, to any other Place; where it is impossible . . . to perform what the World in general reasonably expect from us."[68] Obviously to emphasize that same intention, but in a more tactful tone, Jeremiah Dixon wrote the Society two days later to say that "we shall be very sorry to proceed from this Place, to any other; where the Society . . . can gain no Honour, or we any Reputation; and to go . . . merely for the Premium is an Intention far from our first Design."[69]

But the Council failed to appreciate their fears or lend willing ears to their advice. A proposal which was made from the chair and unanimously adopted at the meeting of 31 January 1761 indicated the great surprise of the Royal Society at the position taken by Mason and Dixon in the light of the fact that they had already received several advances towards their expenses. They were warned:

. . . their refusal to proceed upon this Voyage, after their having so publickly and notoriously ingaged in it . . . [would] be a reproach to the Nation in general, to the Royal Society in particular, and more Especially and fatally to themselves: and after the Crown had been graciously . . . pleased to encourage this undertaking, by a grant of Money . . . and the Lords of the Admiralty, to fit out a Ship of War, on purpose to carry these Gentlemen to Bencoolen, and after the Expectation of this & various other Nations . . . [had] been raised to attend the event of their Voyage; their declining it at this Critical Junction, when it is too late to Supply their places, cannot fail to bring an indelible Scandal upon their Character, and probably end in their utter Ruin.[70]

[67] *ibid.,* fol. 130.
[68] *ibid.*
[69] *ibid.,* fol. 131. Letter to Charles Morton from Jeremiah Dixon dated Plymouth, 27 January 1761.
[70] CMRS, IV, 298-301.

Then, to indicate that the threat was more than verbal, the Council informed them that the Society would, "with the most inflexible Resentment," bring them to court and prosecute them "with the utmost Severity of the Law." Whether or not they feared the French less than an English court, or the distant loneliness of Sumatra less than the inflexible resentment of the Royal Society we shall never know. But whatever the reason, a curt note to the Council from Plymouth dated 3 February 1761, announced that "their dutiful servants" would depart that same evening.[71] The harshness of the Society's admonition was realistically softened, it should be added, in a postscript to the above vote of censure. Since the uncertainties of war and weather remained with them no matter what the circumstances of departure or the precautions at sea, it was left "to the discretion of the Observers when they are upon their Voyage, to act in the best Manner, for Effectually answering the ends of their designation. . . ."[72] That discretionary power was employed early, for when next heard from, Mason wrote that they were preparing to observe the transit of Venus at the Cape of Good Hope, having arrived there 27 April 1761. To this letter was added the bitter postscript that Bencoolen had been taken by the French.[73]

The expeditions which sailed from Britain bound for Bencoolen and St. Helena were not the only ones to hoist the Union Jack in the name of the transit of Venus in 1761. Under the stimulus of John Winthrop, Hollis Professor of Mathematics and Natural Philosophy at Harvard, the Province of Massachusetts sponsored an expedition to St. Johns in Newfoundland. A letter from Winthrop to Francis Bernard, Governor of the Colony, on the importance of the event, led to an appeal delivered to the House of Representatives in the Governor's name on 18 April 1761.[74] Addressed as it was to a group of eminently prac-

[71] Roy. Soc., Misc. MSS, X, 134.
[72] CMRS, IV, 301.
[73] Roy. Soc., Misc. MSS, X, 135. Letter from Mason dated Cape of Good Hope, 6 May 1761.
[74] John Winthrop, A Relation of a Voyage from Boston to Newfoundland, for the Observation of the Transit of Venus, June 6, 1761 (Boston: Eddes & Gill, 1761), p. 22. The Governor's message was printed as an appendix to this pamphlet.

tical men, it is nevertheless interesting to record that it is not very different from Chabert and La Caille's memoir to Louis XV, or the Royal Society's prayer to George II. All of these addresses, Governor Bernard's no less than the rest, offer an insight into contemporary social apologetics, that mixture of private ends with public means in the name of national honor or well-being that so often appears in these appeals. "You must know," the Governor told the House of Representatives, "that this Phenomenon, (which has been observed but once before since the Creation of the World) will, in all Probability, settle some Questions in Astronomy which may ultimately be very serviceable to Navigation: For which Purpose, those Powers that are interested in Navigation, have thought it their Business to send Mathematicians to different Parts of the World to make Observations."[75] Bernard's argument, as Winthrop's tactful letter to him must have suggested, was mainly utilitarian. Nowhere does he suggest the importance of the transit of Venus for obtaining the solar parallax, and therefore the actual dimensions of the universe. For that matter, he does not even inform them of the precise determinations of longitude to be made in connection with the observations, the one item in the entire operation which is really "very serviceable to Navigation," and which might have appealed most strongly to the Representatives. But then perhaps it was not necessary to overplay the argument. The practical case, stated generally, was seconded only by a vague reference to national honor, by mention of the King's interest in the project, the participation of the French and other nationals in parallel undertakings, the "Cause of Science," and the prestige of the Province. Finally, the suggestion was made that the province sloop, very shortly due for a voyage to Penobscot anyway, be used to "proceed from thence with the Professor to Newfoundland."[76]

A committee of five was then appointed to consider the Governor's message and report thereon. Two days later, on 20 April 1761, that report was given and the House voted to fully support Winthrop's ambition and the Governor's wishes. Captain Saunders, "Master of the Sloop in the Province Service," was

[75] *ibid.*, pp. 22-23.   [76] *ibid.*, p. 23.

ordered to wait on Winthrop and convey him "with the Apparatus and other Necessaries," to northeastern Newfoundland or any other place judged suitable for the observations by Winthrop.[77] The necessaries and apparatus so willingly transported by the province sloop were supplied by Harvard University. By a vote of the Corporation on 5 May 1761, Winthrop was invited "to take with him upon his said voyage, such astronomical instruments of our Apparatus, as he thinks he shall need."[78] These consisted of a pendulum clock, a Hadley's octant with Nonius divisions, a refracting telescope with cross hairs, and "a curious reflecting telescope, adjusted with spirit-levels at right angles to each other . . ."[79] (Figure 11). Whether the Harvard Corporation and the Province of Massachusetts meant to include additional personnel in their joint provisions for the expedition or not, Winthrop took their encouragement broadly and invited three men to accompany him as assistants: Williams and Rand, two of his pupils "who had made good proficiency in mathematical studies," and Moses Richardson, the College carpenter.[80] The expedition left Boston on 9 May 1761, and thirteen days later arrived at St. Johns, Newfoundland.

The total effort of the British to observe the transit of Venus of 1761 was thus unexpectedly enlarged by the independent colonial expedition to a new station. For all their lateness on the scene, they were therefore able to match the French in the number of expeditions dispatched overseas to take advantage of the first of the two transit opportunities of the century. The strength of conviction which lay behind the drive to take part in these observations is perhaps best summarized by one of the

[77] *ibid.*, p. 24. The vote of the Massachusetts House of Representatives is printed here as an appendix.

[78] Harvard University Archives, Corporation Records, II, 142, quoted by I. B. Cohen, *Some Early Tools of American Science* (Cambridge: Harvard University Press, 1950), p. 37.

[79] Winthrop, *op.cit.*, pp. 8-9. This telescope, made by Benjamin Martin and now in possession of the Science Museum, South Kensington, London, was a gift of Thomas Hancock to Harvard College in 1761. It was the only instrument known to have survived the Harvard Fire in 1764. See Cohen, *op.cit.*, p. 39.

[80] Winthrop, *op.cit.*, p. 9. Williams and Rand are mentioned by Winthrop in a printed note, to which Moses Richardson's name has been added in a contemporary hand on the copy of the pamphlet to be found in the Houghton Library, Harvard.

leading English instrument makers and scientific popularizers of the day, Benjamin Martin. "If we make the best Use of each [of the transits of Venus of 1761 and 1769]," he wrote, "there is no doubt but Astronomy will, in ten Years Time, attain to its ultimate Perfection."[81]

[81] B. Martin, *Venus in the Sun: Being an Explication of the Rationale of that Great Phaenomenon . . . and of the Manner of applying a Transit of Venus over the Solar Disk, for the Discovery of the Parallax of the Sun . . . the Theory of that Planet's Motion, and . . . the Dimensions of the Solar System* (London: W. Owen, 1761), p. xi.

THE EXPEDITIONS AND THEIR
RESULTS

THE expeditions under French and British auspices which left for distant parts to observe the transit of Venus of 1761 were not of course the only important undertakings of the same sort to take place in the name of the transits. But within the framework of this problem and the compass of their times, they were the most outstanding of all in range, variety, and human interest, and therefore deserve full and careful narration.

The wholesome tendency of the French observers to record the details of their voyages in reminiscent volumes and daily journals is offset only by the formidable length to which they run. Both Chappe d'Auteroche and Le Gentil de la Galaissière published multi-volume editions of their respective Siberian and Oriental adventures, while Pingré has left us a closely written diary of 140 quarto-sized folio sheets, as yet unedited and unpublished. In contrast to this Gallic abundance, our knowledge of the actual voyages of Maskelyne and Waddington, Mason and Dixon, and John Winthrop and his students is almost entirely dependent on a few letters, a pamphlet or two, and the very lean pickings from the official memoirs and transactions of the appropriate scientific societies. Whatever other reasons may be found, say in the personalities of the men in question, the differing cultures of their native lands, or the varied and exotic nature of the geography through which the pursuit of Venus took them, one explanation of this extreme documentary contrast lies in the relative ease with which the British observers were able to reach their destinations and fulfill their ambitions. The brief naval engagement in the English Channel involving Mason and Dixon and the French navy, while intense enough at the time, is almost trivial by comparison with the tribulations

97

of the French observers. And John Winthrop's thirteen-day jaunt to Newfoundland in the spring of the year can in no way be considered a dangerous adventure. But excitement and extreme danger were the frequent fare of the French astronomers. This adventure, in the fullest sense of that word, a deepened anxiety over the scientific aspects of their often-frustrated efforts, and a desire to complain about the slings and arrows of their outrageous fortune are undoubtedly the sources of the bulky French rationale. So it is to this dissatisfaction in the souls of articulate eighteenth-century Frenchmen that we owe the lively narrative of scientific expeditions to capture Venus in transit.

Alexandre-Gui Pingré whom the Académie des sciences selected for its expedition to the Isle Rodrigue in the Indian Ocean, was one of the most interesting men ever to enter the ranks of its membership. Born in Paris in 1711, his first education, and indeed part of his early career, was devoted to theology. By his twenty-fourth year he was a professor of the subject and a member of the Order of Sainte Geneviève, into whose congregation he had been admitted at the age of sixteen, "cedant uniquement à l'impulsion de la sensibilité et à la passion d'apprendre."[1] An eclectic in thought, with a penchant for unconstrained opinion that led even to a sympathy for freemasonry and the composition of masonic verse, Pingré soon fell from grace in the eyes of the Church hierarchy, then so deeply troubled by the heretical rumblings of Jansenism and therefore suspicious of independence in any form. He was removed from his chair in theology, though not from his Order, and relegated to an obscure elementary school, there to bring the rudiments of Latin grammar to the young and the indifferent. But he was to find no peace there. Accused of bringing suspect doctrine to the minds of his innocent charges, he received five *lettres-de-cachet* in the space of four years.[2]

[1] R. de Prony, "Notice sur la vie et les ouvrages d'Alexandre-Gui Pingré, Membre de l'Institut national des sciences et arts," *Mémoires de l'Institut nationale des sciences et arts: Sciences mathématiques et physiques*, 1 (1796), XXVI.
[2] E. P. Ventenat, "Notice sur la vie du citoyen Pingré, lue à la séance publique du Lycée des Arts," *Magasin encyclopédique*, 2e année, Vol. 1 (1806), VII, 344. This was also inserted in the *Mercure* (10 prairial an 4 [1796]), XXII, 217.

1. Joseph-Nicolas Delisle

2. Jean Chappe d'Auteroche

3. Father Maximilian Hell

4. Alexandre-Gui Pingré

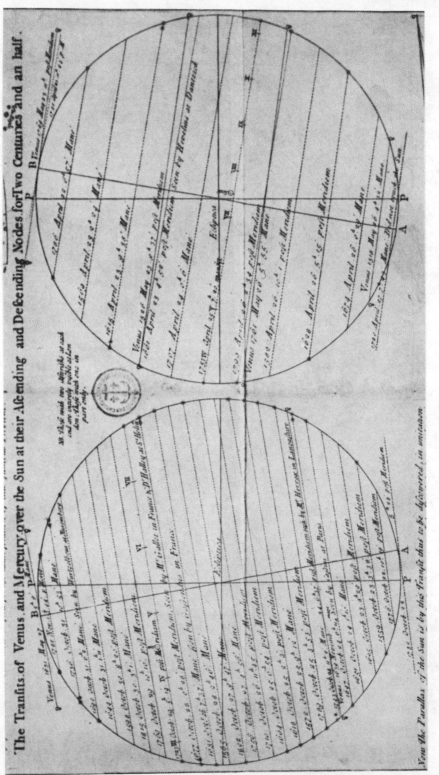

5. The Transits of Venus and Mercury over the Sun

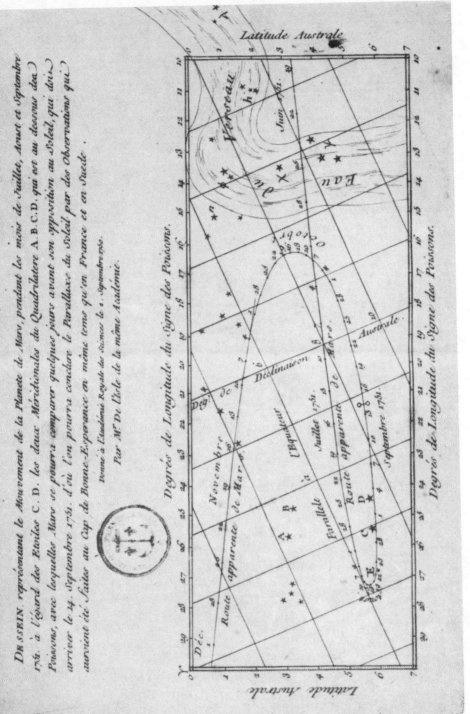

6. The Movement of Mars

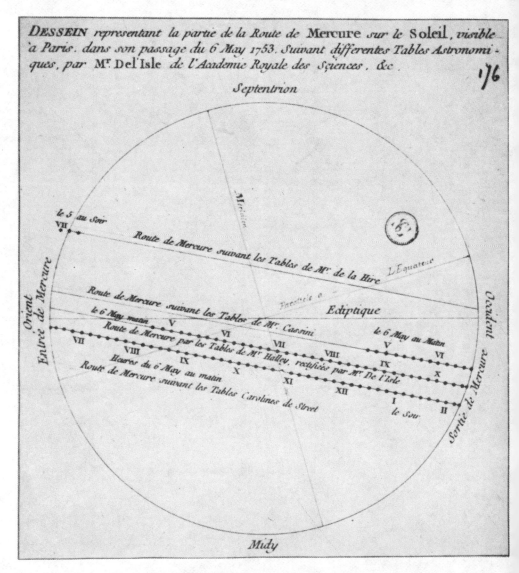

DESSEIN *representant la partie de la Route de* Mercure *sur le* Soleil, *visible à Paris, dans son passage du 6 May 1753. Suivant differentes Tables Astronomiques, par* Mr. DelIsle *de l'Academie Royale des Sçiences, &c.*

176

Septentrion

Meridien

Orient

Entrée de Mercure

le 5 au Soir
VII

Route de Mercure suivant les Tables de Mr. de la Hire

L'Equateur

Route de Mercure suivant les Tables de Mr. Cassini

Parallele a

Ecliptique

le 6 May matin
V

Route de Mercure par les Tables de Mr. Halley, rectifiées par Mr. De l'Isle

le 6 May au Matin

VII

VIII

IX

Heures du 6 May au matin

Route de Mercure suivant les Tables Carolines de Street

V   VI

VII

VIII

IX

X

X

XI

XII

I

le Soir

II

Occulent

Sortie de Mercure

Midy

7. The Path of Mercury across the Sun: Paris, 1753

8. *Mappemonde* for the Transit of 1761

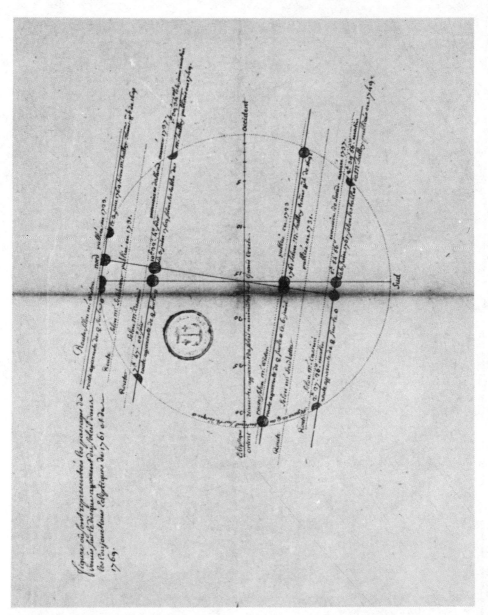

9. The Transits of Venus, 1761 and 1769

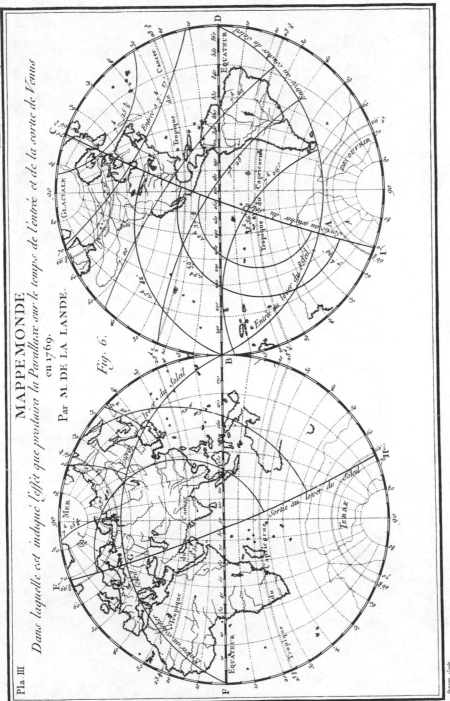

Mem. de l'Ac. R. des sc. 1757. Page 260 Pl. 26.

MAPPEMONDE

Dans laquelle est indiqué l'effet que produira la Parallaxe sur le temps de l'entré et de la sortie de Vénus
en 1769.

Par M. DE LA LANDE.

Fig. 6.

10. Mappemonde for the Transit of 1769

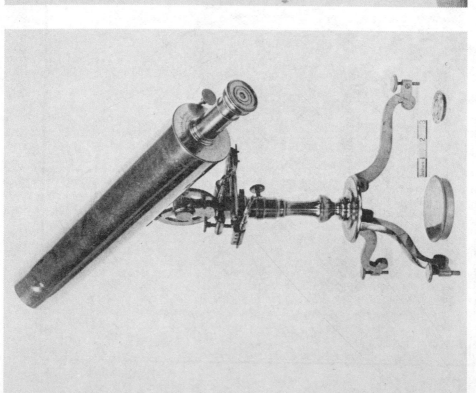

12. Joseph Brown–Benjamin West Reflecting Telescope

11. Gregorian Telescope by Martin

During this period, the famous French surgeon Le Cat founded an academy of sciences at Rouen, for which he lacked an astronomer. Pingré, who had recently settled in Rouen, was thirty-eight when Le Cat approached him with an invitation to fill that need. His acceptance brought him in contact with astronomy for the first time; nevertheless a calculation of the lunar eclipse for that same year, 1749, was good enough to put him in touch with the Académie des sciences at Paris. Elected a correspondent of the Académie, his growing scientific fame, strangely enough, helped to restore his reputation in the Church, and he was named a librarian of Sainte Geneviève and returned to Paris, where the Order erected a small observatory on its abbey for his use.[3] Not long thereafter, he was given the rank of *associé libre* in the Académie, the highest rank permitted to a churchman, according to a regulation dating from 3 January 1716.[4]

In the years before his voyage to Rodrigue, Pingré was closely connected with Le Monnier and calculated a kind of nautical almanac, *l'Etat du Ciel* (1754-1757), designed to complement the *Connoissance des Temps*, which had been published since 1682. He was to make several other voyages after the Rodrigue expedition, mainly for the purpose of testing the marine clocks of Pierre Leroy and Ferdinand Berthoud: in 1767 with Messier, in 1769 with Fleurieu, and in 1771 with Verdun and Borda.[5] His most important published work was the *Comètographie* (Paris: 1783), an historical and theoretical treatment of comets that was definitive for its day. Interestingly enough, he thought that comets might some day serve as another kind of natural phenomenon for obtaining the solar distance. The variety of

[3] Delambre, "Alexandre-Gui Pingré," *Biographie Universelle*, xxxiii, 364.

[4] A. Lacroix, *Notice historique sur les membres et correspondants de l'Académie des sciences ayant travaillé dans les colonies Françaises des Mascareignes et de Madagascar au XVIIIe siècle et au début du XIXe* (Paris: Gauthier-Villars, 1934), p. 24.

[5] Delambre, "A.-G. Pingré," *op.cit.*, pp. 364-365. For valuable details on these voyages, much can be obtained from two excellent studies: Jean Mascart's *La Vie et les travaux du chevalier Jean-Charles de Borda (1733-1799), épisodes de la vie scientifique au XVIIIe siècle* (Lyon: A. Rey, 1919), which is a veritable gold mine of information on eighteenth-century science; and F. Marguet, *Histoire de la longitude à la mer au XVIIIe siècle, en France* (Paris: A. Challamet, 1917).

his interests was amazingly broad and surprisingly deep. A two-volume translation of the Latin astronomical poem of Manilius, published in 1786, earned the unstinted praise of no less a poet and classicist than A. E. Housman as "the best and far the best existing text of Manilius."[6] To Pingré as astronomer, geographer, theologian, freemason, one must therefore add a great ability as Latinist.

But his published works are as nothing compared to the mountain of unedited manuscripts on deposit at the Bibliothèque Sainte Geneviève in Paris: translations of early Spanish voyages, history and historical criticism, a discourse on Chinese astronomy and chronology, musical satires and liturgical hymns, a treatise on lotteries, literary sketches and portraits from Ovid to his contemporaries, and poetry everywhere, of every kind, in French and Latin, epigrams, sonnets, rondeaus, elegies, and songs. The inexhaustible Pingré easily exhausts his reader, and we must interrupt the enumeration in order to proceed to the Isle Rodrigue.[7]

Immediately after the decision to send Pingré to Rodrigue in November 1760, the Académie must have written directly to Britain, in what is certainly a most unusual wartime solicitation between belligerents. This was the request for a *laissez-passer* from the British Admiralty to Pingré, to grant him safe and uninterrupted passage to and from Rodrigue. The petition was honored, and on 25 November 1760 a general order addressed to all commanders of British vessels was dispatched to Pingré containing instructions *"not to molest his person or Ef-*

---

[6] A. E. Housman, M. *Manilii Astronomicon* (London: Grant Richards, 1903), I, xx. "In 1786 appeared at Paris . . . the text of A. G. Pingré, with a French translation or rather paraphrase facing it, and a frugal equipment of brief notes textual and explanatory. Pingré, though intelligent and well read, was no marvel of learning or brilliancy or penetration; but the prime virtue of a critic, worth all the rest, he had: simplicity and rectitude of judgement. The text . . . is yet even now, in the year 1903, the best and far the best existing text of Manilius. Pingré's own conjectures are not many, but sensible and sometimes excellent. . . . In no edition of Manilius is there so little which calls for censure."

[7] No attempt can possibly be made to enumerate the manuscript material pertaining to Pingré at the Bibliothèque Ste. Geneviève, but those documents which are of importance to the transits of Venus are: Pingré MSS, 531, 533, 1085, 1175, 1803, 1805, 1977, 2312, 2321, 2342, and 2551. See also Lacroix, *op.cit.*, 25-27.

fects upon any account, but to suffer him to proceed without delay or Interruption."[8] (See Appendix III, D, and note.)

Pingré left Paris on 17 November 1760, after a big farewell dinner the evening before with many friends, including La Caille. Though the good cheer flowed and the conversation ran well into the night, Pingré tells us that he dined without appetite, terrified by the forthcoming voyage and fearful that he might well be seeing his friends for the last time.[9] The trip to the coast of Brittany by carriage, via Alençon and Rennes, was very tiring. From a little village near Rennes he sent his first letter to De Fouchy, Secretary of the Académie, within a week after his departure, to tell him that all went well with himself and most of the instruments, "mais au diable si l'on peut transporter un Barometre sans lésion quelquonque."[10] This problem of safeguarding the instruments was one he was to face in even greater degree after the ocean voyage. On 30 November he met Denis Thuillier, his assistant, at the little town of Hennebont, not far from Port Louis, L'Orient (written today as Lorient), port of embarkation for the Compagnie des Indes, where they were to board ship. It had taken thirteen days to make the trip from Paris to the coast, and almost every league's worth of detail and incident are noted in Pingré's journal: the dry bread at one place, the quality of woodcock and wine at another, and the all-too-sudden intimacies of companions in the coach. The meeting with Thuillier also led Pingré to reflect with appreciation that the moving forces behind the very idea of an assistant were

[8] Bibliothèque Ste. Geneviève (hereafter cited: Bib. Ste. Gen.), Pingré MSS, 533, No. 2, fol. 85. There is a striking error in this Admiralty document: Isle Rodrigue is placed in the East Indies, rather than in the Indian Ocean east of Madagascar!

[9] Bib. Ste. Gen., Pingré MSS, 1803, Relation de mon Voyage de Paris à l'isle Rodrigue (hereafter cited: Relation), fol. 1. There are two copies of this manuscript, possibly a third. The first and most complete, which I have used, is in Pingré's own handwriting. An abridged and somewhat edited version, handwritten but not by Pingré, is at the Bibliothèque de la Service Hydrographique de la Marine. The third possibility is based on a catalogue reference to the Bibliothèque de la Port de Brest. I have a feeling that this third reference is really to the same copy at the Service Hydrographique, which was the recipient of much of the documentary material concerning naval affairs that was moved from Brest to Paris.

[10] Archives de l'Académie des Sciences (hereafter cited: Arch. Acad. Sc.), Dossier Pingré, Lettre de Pingré à Grandjean de Fouchy, 23 Novembre 1760, fol. 1.

Le Monnier, Cardinal de Luynes, and the King's minister, St. Florentin.[11]

Throughout December the impatient voyagers stayed in Lorient, while the *Comte d'Argenson*, which was to carry them to Rodrigue, took on cargo and prepared for sailing. Originally constructed as a man-of-war with a fire power of sixty-four guns, the *Comte d'Argenson* had had its armament reduced to twenty-six pieces in order to increase its carrying capacity to about a thousand tons.[12] It was this limited space, not atypical of the times, which led to heated arguments between Pingré and the company's agent at the port over the quantity of their baggage. Seven to eight hundred pounds, Pingré argued, were not too much for an astronomer! Eventually the conflict was resolved, but not before it had been carried back to the Académie des sciences. The letters between Pingré and Buffon that settled the issue also brought money to the voyagers; hence, they indicate some of the expenses involved and, at the same time, give us an opportunity to overhear a note of potential discord between Thuillier and Pingré concerning an incidental feature of the voyage. "M. de Buffon," wrote Pingré, "m'envoyoit . . . 50 pistoles que la Comité de la Trésorerie m'ajugeoit sur un Mémoire de frais que j'avois presenté: il avoit 10 louis à M. Thuillier pour des commissions d'animaux &c. Je n'ai point vu ces 10 louis: ainsi je ne suis chargé de rien."[13]

Above and beyond Pingré's obvious annoyance at the direct financial transaction between Buffon and Thuillier, in view of his own superior rank in the expedition, the brief note is very informative. The connection between Buffon and natural history needs no elaboration here, but this action does show a secondary use to which these astronomical expeditions were sometimes put. It seems certain that Thuillier was asked to collect specimens of new or interesting flora and fauna for Buffon's study or for the *Jardin du Roi*. This project was eventually fulfilled with Pingré's complete cooperation for his anger was not to persist. Unfortunately, much of the collection was after-

[11] *ibid.*, fol. 3.
[12] J. Vivielle, "Les tribulations d'un astronome dans la mer des Indes," *Communications et mémoires de l'Académie de Marine*, IV, 2 (1925), 12.
[13] Pingré, *Relation*, fol. 3 verso.

ward dispersed. A detailed description of the plant and animal life of Rodrigue and its surroundings was prepared, which today is of very great interest because most of the area's ecology has radically changed since then.[14]

By 9 January 1761 all was ready, and with a full cargo and some sixteen passengers, the *Comte d'Argenson* weighed anchor in the late afternoon. Of Pingré's fellow passengers, about twelve were new employees of the Compagnie des Indes going to seek their fortune in the Orient. Others included M. de Saint-Jean Estoupeau, heading for the second-in-command at Pondichery but destined never to reach his post, and a Monseigneur Edme Bennetat, Bishop and Apostolic Vicar of Indochina. In command of the ship was Dufresne Marion, captain in the French navy and explorer, whose name, together with that of his lieutenant Crozet, also aboard the ship, would one day be given to a group of islands in the southern oceans.[15]

As in the case of Mason and Dixon, misadventures began early for Pingré and Thuillier. Only one day out from the coast of Brittany, after a night about which Pingré wrote, "J'ai peu dormi; la goutte me tormente encore," they sighted a fleet of five English ships. The captain's call to quarters required most of the partitions dividing the passengers' quarters to be torn down in order to free the cannon, with the result that luggage and passengers alike were all tossed into Pingré's quarters, which were apparently spared this alteration. A ship's council was held and the decision made to return to port if possible rather than risk an encounter with the British. But a fortunate shift in the wind, Marion's skillful maneuvering, and the long winter night made it possible for them to elude the enemy, and by 12 January 1761 they were truly en route for Rodrigue.[16]

Good weather and good speed made for good humor aboard ship, and often there were singing and dancing to the sound of the Chevalier de Mouy's violin for Pingré and the officers, and to the swirl of bagpipes for the crew. As with all long sea voyages, much time was spent in playing cards, in philosophi-

[14] See Lacroix, *op.cit.*, p. 27.
[15] Pingré, *Relation*, fol. 2 verso, and Vivielle, *op.cit.*, p. 12. Like Captain Cook, Marion was murdered by natives, in his case of New Zealand (in 1772).
[16] Pingré, *Relation*, fol. 4.

cal discussions on the wind and weather, the phosphorescence
of the sea, which Pingré tried to capture for examination under
a magnifying glass, and in watching the officers demonstrate "le
noble jeu de la savate." But throughout January they were still
zigzagging to avoid encounters with enemy fleets and fleeing
at the sight of unknown sails in the distance.[17]

The voyage also gave Pingré an opportunity to experiment
with the calculation of the longitude at sea, the most important navigational problem of the eighteenth century. Following
the method which La Caille had carefully developed in the
course of his voyage to the Cape of Good Hope earlier in the
century, Pingré, Thuillier, and the officers of the crew made
regular observations of the angular distances of the moon, from
either the sun or certain fixed stars. By means of these data and
a series of laborious but not too-difficult calculations in which
various sets of lunar tables were used, it was possible to determine the longitude at sea. One incident in the voyage involving
this technique of lunar distances is highly illustrative of both
the contemporary state of maritime cartography and the difficulty of determining the longitude at sea before the use of
chronometers.

By 29 January 1761 the *Comte d'Argenson* was somewhere
in the mid-Atlantic Ocean, west of Dakar on the African coast.
On that date, Pingré observed the altitude of the sun, while
Crozet observed the angular distance between the eastern edges
of the sun and moon. The calculation which Pingré made from
these observations placed them at 22 degrees, 32 minutes west
of the Paris meridian. This meant that the Cape Verde Islands
which they were fast approaching would all lie completely west
of their route, because the easternmost tip of the most eastward island, Bonavista, was at about 25 degrees west longitude
from Paris. Yet the assumption of the ship's captain and his
officers had been that they would pass west of the islands.[18]
Neither party swore upon the absolute certainty of his argument, and the cautious captain hove to at nightfall to avoid
the possibility of running into the islands during the night.

17 *ibid.*, fol. 8.
18 *ibid.*, fol. 10 verso.

The next day, 30 January, at six in the morning, they saw the southern part of the island of Santiago some six or seven leagues to the west. According to the chart of the coastal waters of West Africa (dated 1753), which the Départment de la Marine had issued to them, they should then be running right into the island of Bonavista, but the way ahead was all clear, and Bonavista not to be seen. Pingré's conclusion was that they had passed East of the Cape Verde Islands and that the longitude of Bonavista had been incorrectly given on the map by as much as 2 degrees.[19] The days which followed this incident were filled with debates on the virtues of various methods for calculating the longitude at sea. Marion objected to the length and difficulty of Pingré's calculations, believing that the method of lunar distances would yield no greater accuracy than eighty nautical leagues and that furthermore, "ils étoient au-dessus des marins." In this latter point he was probably quite right, but Pingré promised that "si la Lune paroissoit 8 ou 10 jours avant que nous doublessions le cap de bonne Espérance, nous en don-nerions la longitude à 12 ou 15 lieues près. . . ."[20] Yet even the most experienced astronomical computer was at the mercy of the inaccuracy inherent in the tables utilized, and though the method of lunar distances was widely accepted in the second half of the eighteenth century, its operational precision was far from satisfactory.

The ever-present ennui of such a long voyage was probably han-dled better by Pingré than anyone else. The daily routine of astro-nomical observation and calculations which he set himself, the desire to record all the petty anecdotes of shipboard life, and the urge to describe new natural phenomena (or at least that which was new to him) filled his days. But often these more serious occupations were not enough, and much surplus energy was put into the traditional King Neptune ceremony, known to the French as *le Pere de la Ligne*, at the crossing of the

[19] *ibid.*, fol. 12.
[20] *ibid.*, fol. 14. On the use of this method in the eighteenth century, see J. B. Hewson, A *History of the Practice of Navigation* (Glasgow: Brown, Son & Ferguson, Ltd., 1951), pp. 231-238. A better account of its limitations is given in F. Marguet, *Histoire générale de la navigation du XVe au XXe siècle* (Paris: Soc. d'Editions Géog., Maritimes et Coloniales, 1931), 235-247.

equator, and into a variety of practical jokes that included capturing sea birds and banding them with *épigraphes,* of which the following is a relatively polite example: "Cocu soit le premier qui me prendra."[21] When the seas were rough, alcohol was a friend in need, and even of direct assistance in astronomical observation. "Des liqueurs," he writes, "nous donnent la force nécéssaire pour prendre une distance de la Lune au Soleil."[22] And one day after a rather long interval of such a build-up of the necessary force for what he called "la prise Astronomico-Bacchique de quelques hauteurs," he was proud enough to write that "j'ai pris hauteur, non pas avec la bouteille, mais avec l'Octant."[23]

Perhaps plagued by too many astronomical conversations on Venus, one of Pingré's fellow passengers ("le Prophete de malheur" Pingré called him), penned an epitaph for Pingré, whose allotted time, he felt, would run out at Isle Rodrigue.

> Ci-gît qui cherit tant Vénus
> Qu'à Rodrigues il fut surprendre:
> De l'astrologue *in partibus*
> Cher Passant, respecte la cendre.[24]

The good disposition with which Pingré usually faced things did not survive this dry humor. For while he does enter the epitaph in his journal, he also swears that, full of life, he will be found at Paris in the years ahead, ready to receive congratulations on the successful conclusion of the voyage, even from "Monsieur le Prophete." Actually, he was to do much better than that, for he survived this voyage, three others, and the Revolution, dying at the age of eighty-five in the arms of Morpheus rather than of Venus.

By April the *Comte d'Argenson* had rounded the Cape, and Pingré felt that the worst part of the voyage was over. In keeping with the custom of the day, a solemn *Te Deum* was sung, "au son de toutes les cloches de navire," by everyone aboard. "Tous chantoient," he goes on to say, "de plein coeur, et à plein gosier . . . celui ci chantoit à la Parisienne, l'autre à la Romaine

---

[21] Pingré, *Relation*, fols. 20-23. [22] *ibid.*, fol. 27.
[23] *ibid.*, fol. 32 and fol. 29 verso. [24] *ibid.*, fol. 33 verso.

... cela formoit la musique la plus singulière du monde."[25] But the optimism inspired by rounding the Cape and the singing of the extraordinary *Te Deum* was soon dispelled. Within a few days they met and successfully eluded enemy ships. But the sense of security was gone, for the Indian Ocean was far from calm during the Seven Years' War. On 8 April they picked up *le Lys*, the last of a six-ship French group sent from the Mascarene Islands to the Cape of Good Hope to obtain supplies. About ten days out from the Cape, she too had been pursued by the British, and now was taking water badly. Requested to stand by, the *Comte d'Argenson* cut canvas and slowed down.

At first this pleased Pingré, for it meant company on the lonely ocean, relatively recent news, and fresh meat, fruit, and wine from the *Lys's* newly acquired stores. The news was compiled from reports turned in at the Cape of Good Hope by Danish and Dutch ships out of Tranquebar and Bengal. Conflicting reports about the British siege of Pondichery were delivered, with nothing conclusive on the matter to allay any doubts one way or another. But the news that Bencoolen had been taken by the French some time between July and October of 1760 was joyously received.[26] Little did Pingré know that about two weeks later Mason and Dixon would arrive at the Cape to discover this French victory that would prevent an observation at that important station.[27]

But the rapid change in physical circumstances and the relatively good cheer brought about by the chance encounter between the two French ships on the high seas, so immediately and enthusiastically welcomed, soon turned into anger, ill-will, and frustration. When Pingré learned that Captain Blain des Cormiers of the *Lys*, senior to Dufresne Marion and therefore capable of giving the order, had asked that the *Comte d'Argenson* accompany the *Lys* almost indefinitely, he became furious. For this meant that there would be no landing at Isle Rodrigue. "Je ne pouvois m'imaginer," he angrily wrote on 12 April, "que le Roi m'est envoyé pour être à la merci des Marion & des Blain:

---

[25] *ibid.*, fol. 38.   [26] *ibid.*, fol. 43 verso.   [27] See Chapter ii.

ils ont sans doute leurs raisons: mais, j'ai droit sans doute aussi d'avoir les miennes."[28] Under pressure from Pingré, Marion then asked Blain to allow him to go on at his more rapid pace in order to fulfill his obligation to the astronomers. The response to this request was a written order from Blain supporting his earlier request. Caught thus between Blain's command and Pingré's anger, Marion defended himself as best he could, pointing out to Pingré that his original orders did not include any directly from the King. And furthermore, those that were addressed to him from the Compagnie des Indes simply asked him to deposit Pingré and Thuillier at Isle Rodrigue if, in his opinion, it appeared safe to do so. In the heat of the argument, it was even suggested by one of the officers, as Pingré records it, "qu'il faloit nous jeter dans la mer" to solve the problem.[29]

This kind of frustration, only a short distance from his goal, was almost more than Pingré could bear, and he drew up a formal letter of protest to Marion, which Thuillier also signed. (See Appendix iii, A–B for both the Pingré letter and Marion's reply.) In the rolling prose of eighteenth-century eloquence, Pingré reviewed the entire case in terms of the King's orders and those of the syndics of the Compagnie des Indes, the wishes of the Académie des sciences, and the hopes of all Europe, whose eyes were carefully focused upon the play of events in the Indian Ocean. For these reasons, Pingré wrote, "Nous vous sommons, Messieurs, par le présent act . . . de suivre le plan de route, vous vous étiez proposé jusqu'à la rencontre du vaisseau, le Lys; afin que nous puissions debarquer à temps à l'isle Rodrigue. Nous protestons formellement contre tous les arrangements, fondées peut-être d'ailleurs sur des faits hazardés, et sur des conjectures hors de vraisemblance, seroient contraires aux interets de notre commission, et aux ordres respectables, dont nous sommes . . . chargés. Enfin nous croyons autorisés à vous demander une réponse claire, catégorique et par écrit à notre sommation présente. . . ."[30] The reply which Pingré received to this "pièce d'éloquence" was immediate and to the

[28] Pingré, *Relation*, fol. 44.   [29] *ibid.*, fol. 44 verso.
[30] *ibid.*, fol. 45. This is of course a copy of the actual letter which Pingré wrote into the journal.

point. Marion informed the astronomers that, so far as he was concerned, they were still directly on the route to Rodrigue, that he was subordinate to the Captain of the *Lys*, responsible to the Governor of the Isle de France and the members of its Council, and that if this reply was unsatisfactory they could appeal to his immediate superior in the area, the same Captain Blain des Cormiers of the *Lys*.[31] The obvious logic of Marion's position, delivered bluntly with a derogatory remark about Thuillier's manners, only increased the tension aboard ship. But not for long; within a few days Marion and Pingré dined together, and aided by the liberal flow of the newly supplied "vin blanc du Cap," sweet reason was once again restored.

By 6 May 1761 the astronomers had landed at Isle de France, and three days later they were aboard the corvette *la Mignonne*, on their way to Rodrigue. During the brief interval on Isle de France, Pingré had arranged to have samples of animal life native to the island group sent back to Buffon for the *Jardin du Roi*.[32] Although the distance between Isle de France and Rodrigue is only about 320 miles, squalls and rough seas prolonged the voyage to nineteen days, so that Pingré and Thuillier did not debark at Rodrigue until 28 May 1761. At that, they had had a real moment of anxiety when, after having sighted the island on 26 May, they were practically becalmed for two days. "Il faut se résigner," Pingré wrote, "à la providence de Dieu et à la prudence de M. des Moulières," the Captain of the *Mignonne*. And the next day, though certain they would not reach Rodrigue then, Pingré optimistically wrote: "Le calm continue sur la mer, dans l'air, et dans l'esprit de M. Thuillier mais non pas certainement dans le mien. . . ."[33] The spirit of this "chanoine regulier" prevailed, and when the *Mignonne* anchored alongside of a sistership, the *Volant*, which had preceded them from Isle de France to collect a cargo of turtles, the astronomers had nine days to prepare for the rendezvous with Venus in transit. Incidentally, turtles were a staple in the islands, and Pingré declared "foie de tortue" to be incomparable to all others.

[31] *ibid.*, fol. 45 verso, and Appendix 1,C.
[32] *ibid.*, fols. 53-54.
[33] *ibid.*, fol. 60 verso.

These were feverish days in which the instruments were unloaded and the island explored to locate the best site for the observatory. Fringed by coral reefs, Rodrigue is an island of basaltic rock rising abruptly from the sea. Something of the physical difficulty which they must have faced in embarking and installing their equipment can be gleaned from the report of the British expedition sent to Rodrigue to observe the transit of Venus in 1874, when there was modern equipment to assist in the task. "The disembarkation of the huts, instruments and other heavy stores," wrote one of the crew in this nineteenth-century expedition, "was necessarily a tedious operation, and could only be affected at high water."[34]

Aided by the officers and crew of the *Volant* and the *Mignonne*, Pingré and Thuillier were able to construct only a very crude shelter for the instruments in the little time left them. Pingré's journal, ordinarily so verbose, is strangely silent on the observatory at Rodrigue, but the official memoir, which he presented to the Académie years later, tells us a little of what it must have been like.

Je me suis contenté de faire placer en plein air quatre pierres assez grandes, & passablement unies; cet endroit qui avoit tout l'air d'être destiné à placer un jeu de quilles, devoit servir à prendre avec le quart-de-cercle les hauteurs des Astres; je fis élever à côté deux espèces de mâts: à l'aide de poulies & de grosses ficelles, j'y appuyois mes lunettes, lorsqu'elles devenoient nécessaires aux observations. Dans une chambre à côté, j'avois placé deux pendules, & cette méme chambre étoit en méme-temps l'unique lieu où je pusse dépôser mes instrumens, encore n'y étoient-ils pas trop à l'abri du vent, de la poussière, & des insultes des animaux. . . .[35]

In spite of careful packaging, the many months at sea had done the instruments little good, and though none were broken, most of the engraved metallic parts were "mangées par la rouille." The olive oil normally used to treat such a condition was lacking, and Pingré improvised by using oil extracted from the ubiq-

[34] G. B. Airy, *Account of Observations of the Transit of Venus 1874, December 8, Made Under the Authority of the British Government: and of the Reduction of the Observations* (London: Her Majesty's Stationery Office, 1881), Part III, Section I, 351.
[35] Pingré, "Observations astronomiques pour la determination de la parallaxe du Soleil, faites en l'Isle Rodrigue," *Mémoires*, 1761, p. 414.

uitous island turtles. By 3 June 1761 the clocks and telescopes were in working order, and between that date and the transit on the sixth, they were able to check their instruments by nightly observations of that great natural timepiece, the planet Jupiter and its satellites. In this case they observed the immersions of the first, second, and fourth satellites.[36]

It rained during the morning of 6 June 1761 at Rodrigue, and afterwards the sun was covered with thick clouds. The immediate sense of frustration which Pingré and Thuillier felt must have been very great indeed, but happily the skies cleared somewhat. Nevertheless they were unable to determine the moments of first and second contact between Venus and the sun because the weather turned bad again at the conclusion of the transit. But they were able to make some useful observations. Pingré made successive measurements of the distances between Venus and the nearest edges of the sun with an 18-foot telescope to which a micrometer had been attached. The time when these measurements were made was also noted. Though not the best, these measurements could be used in a trigonometric technique for determining the solar parallax. Thuillier, in the meantime, observed the transit of the edges of both Venus and the sun across the vertical and horizontal cross hairs of his quadrant.[37] For all their limitations, Pingré considered their observations a success, and that evening at dinner there was much celebrating, with many toasts to all connected with the astronomical mission, from the King and his ministers down to "MM. les Astronomes de tous les pays qui auront observé . . . aujourd'hui Vénus. . . ."[38]

In the days that followed, Pingré and Thuillier determined the latitude and longitude of their station, and together with

[36] Pingré, *Relation*, fols. 61-62 verso.

[37] Pingré, "Observations astronomiques . . . faites en l'Isle Rodrigue," *Mémoires*, pp. 415-482. Between these pages Pingré also discusses the latitude and longitude of Rodrigue, the calculation of the longitude of several other places where the transit had been observed, the transit at Rodrigue itself and various techniques for calculating the solar parallax from a transit: by the method of durations, by interior contact at egress, and by the observation of the shortest distances between the centers of Venus and the sun. Thuillier's observations, it should be pointed out, were of use in determining the diameter of Venus as well as the solar parallax.

[38] Pingré, *Relation*, fol. 62 verso.

the officers of the *Mignonne*, set about to explore Rodrigue and its adjoining islands. They triangulated the high points of the island and those of its neighbors, collected specimens of unusual flora and fauna, and recorded descriptions of a fundamental and primitive biological balance now, alas, forever gone. Here was another demonstration of the eighteenth-century ideal of man in the activities of Alexandre-Gui Pingré. An astronomical expedition was broadened and deepened into a voyage of discovery that somehow related the crabs and turtles of an island, its cartographic image, and the phosphorescence of the sea to the spinning planets and the size of the universe.

These explorations continued through the month of June, during which time the *Volant*, having completed its cargo of turtles, left for Isle de France. The interlude of quiet research and gentle peace, however, was not to last. On 26 June the corvette *l'Oiseau* arrived at Rodrigue, largely, it seems, to enable its Captain to marry the daughter of the island's governor. But three days later the issues of war were brought home to the celebrating islanders, when a British man-of-war, the *Calapate* or *Plassey*, for it seems to have had two names, bombarded the settlement. The brief battle was a complete success for the English. They captured and sacked the island, took the *Mignonne* as a prize, and burned the *Oiseau*. The British passport which Pingré carried was completely ignored by Robert Fletcher, the Captain of the raiding vessel, who apparently treated them most uncivilly.[39]

When the British raiders departed, they left the island with a greater population than it had ever had before, for the officers and part of the crews of both French vessels were left ashore. The enforced imprisonment and isolation lasted about a hundred days, until 6 September 1761, when the *Volant*, making its regular run from the Isle de France, released them from their captivity and brought them back to the larger island in place of its normal cargo of turtles.[40] During the long wait, in which they were "reduits à la seule boisson ignoble de l'eau" to wash

[39] *ibid.*, fols. 67-68.
[40] *ibid.*, fol. 83 verso. Actually the *Volant* arrived on 6 September 1761 and sailed from Rodrigue on 8 September, *ibid.*, fol. 84 verso.

down the rice and flour which had escaped the expropriating activities of Captain Fletcher, they had suffered another raid by two British men-of-war that inadvertantly wandered into the anchorage at Rodrigue. These were the *Drake* and a former French ship *la Baleine*, but this time they received greater courtesy and some provisions. Indeed, Pingré sent some letters to the Royal Society via the Captain of one of these ships. Later on, on 15 September 1761, from Isle de France, Pingré addressed a long letter of protest to the British Admiralty, repeating the narrative of his unfortunate days on Rodrigue, complaining of the violation of his passport and "la conduite illégitime et presque cruelle du Sieur Robert Fletcher," and demanding that justice be done.[41]

From Isle de France, he also wrote Lord Macclesfield, president of the Royal Society, to repeat the tale of Rodrigue and his charges against the British Captain. To the same letter he attached a copy of his protest to the Admiralty, and prayed that some use would be made of it all to punish the Captain for his outrageous conduct.[42] The same day that he wrote to Macclesfield, Pingré sent the Académie des sciences his observations, with the comment, "Elle peuvent ne pas valoir grand'chose," which of course was true, since he had missed seeing the important contacts.[43]

It took our long-suffering astronomers about a month at Isle de France to recover from the ordeal at Rodrigue and prepare for the long voyage home. They sailed from the former island on 17 October 1761 aboard the five-hundred ton frigate *le Boutin*. On the eighteenth they were at Isle de Bourbon, where they spent two months exploring the island and its surroundings,

[41] Pingré, "Lettre à My Lords, les Commissaires préposés pour l'exécution de l'Office du Lord Haut-Admiral de la grande Bretagne et du Irlande &c.," Bib. Ste. Gen., Pingré MSS, 1977, fol. 22 verso. According to Pingré, additional copies of this letter were made: one for the Académie des sciences, to be dispatched from Isle de France aboard the *Adour*, another to be deposited with the Civil Registry at Isle de France and a third which he would carry upon leaving the island; Pingré, *Relation*, fol. 87 verso. (A copy of this wonderful letter is in Appendix III C.)

[42] Arch. Acad. Sc., Dossier Pingré, No. 2, 19 September 1761, Copie d'une lettre de Pingré à Milord Macclesfield.

[43] *ibid.*, No. 4, 19 September 1761, Copie d'une lettre de Pingré à Monsieur le Sécretaire de l'Académie des sciences, fol. 1. The Académie learned of all this on 30 January 1762, when Pingré's letter to De Fouchy was read; *Proc. Verb.*, 1762, fol. 10.

continuing their detailed description of the flora and fauna, and collecting specimens of both whenever they could. Finally, under the stimulus of a fresh Northwest wind, the *Boutin* slipped from her moorings at eleven o'clock at night, 20 November 1761;[44] the last part of the journey was under way. "MM. Les Anglois," Pingré piously prayed in his journal, "je vous estime beaucoup: mais laissez nous tranquilles. . . . Nous avons de si aimables Dames; pour l'amour d'elles encore une fois ne vous trouvez pas sur notre chemin."[45] Unfortunately the English were everywhere, and they found Pingré again.

Save for the petty discomforts of the sea and the interference with his sleep and his calculations by the children of other passengers, the voyage between November 1761 and February 1762 was fairly calm and uneventful for Pingré. But on 11 February 1762, the ominous maxim "l'homme propose et Dieu dispose," marks the beginning of his journal entry for that day.[46] Pursued since dawn by a British ship (called *la Blonde*, according to Pingré), the inevitable combat took place at about two-thirty in the afternoon. It was over in no time with only a little damage to both ships, for the French Captain, upon discovering that he was outgunned, struck his colors hastily, apparently against the wishes of his officers and crew. Once again the British took over, this time with all the good manners of eighteenth-century gentlemen at war. Nevertheless, the passengers saw their baggage broken into and removed, and many of them were transferred to the English ship, as were the French officers.[47] But Pingré and Thuillier were allowed to remain aboard the *Boutin*. On 23 February they sailed into Lisbon. For the wandering astronomers the sea voyage was over.

In the remaining February days, Pingré and Thuillier worked hard to preserve their instruments, their natural history collections, and indeed their entire baggage from successive raids by everyone from the British Captain down to the port stevedores. It was a losing battle for them in every way. The fine collection of shells, sea urchins, parrots, and monkeys (living and dead) which they had carefully put together in the islands was quickly

---

[44] Pingré, *Relation*, fol. 103.   [45] *ibid.*, fol. 103 verso.
[46] *ibid.*, fol. 121.   [47] *ibid.*, fols. 121 verso-122.

dispersed. Even the instruments were in danger. As Pingré indignantly described it in one case: "Le Sr. More a mis ses mains profanes jusques dans le rouage de mes pendules."[48] With the aid of the French Envoy Extraordinary at Lisbon, and the hospitality of friends and fellow churchmen, particularly the French Capuchins in this city, Pingré was able to save the instruments and send them to Le Havre by boat, though they were in a very poor state when dispatched.

Disenchanted by ocean voyages themselves, the astronomers decided to return to Paris overland, and on 28 March 1762 they left Lisbon in a small convoy of carriages and post chaises.[49] Across the bad roads of Spain by day and in worse inns at night, via Madrid and Escurial, often at a speed no greater than the imperturbable gait of a team of oxen, it took them until 26 April to reach Pamplona in the valley formed by the junction of the Cantabrian Mountains with the Pyrenees. On the other side lay France, and two days later they reached the little mountain village of Saint Jean Pied de Port and the road to Bayonne. "Ainsi nous sommes rentrés en France," Pingré wrote with obvious satisfaction, "1 an, 3 mois, 18 jours, 19 heures 53 minutes et demie après l'avoir quittée."[50] Paris was reached the last week of May, not too late for the weary travelers to rediscover the city of light in its springtime magic—a fitting welcome home from the long odyssey.

II

Little is known of the man who volunteered to undertake the arduous journey to Siberia to observe the transit of 1761, in response to the invitation of the Imperial Academy of Science of St. Petersburg which was read at the meeting of the Académie des sciences of 19 April 1760.[51] Only quite recently had Jean Chappe d'Auteroche come to the attention of the Académie, and been admitted to its ranks as *adjoint* upon the promotion of Jérôme de Lalande to *associé* in January of 1759.[52] A student

[48] *ibid.*, fol. 124 verso.   [49] *ibid.*, fol. 126 verso.
[50] *ibid.*, fol. 133.   [51] See Chapter II.
[52] J. P. G. de Fouchy, "Eloge de M. l'Abbé Chappe," *Histoire*, 1769, p. 164. Delambre's discussion of Chappe in his *Histoire de l'astronomie au XVIIIe siècle*, pp. 621-623, adds little to De Fouchy's *éloge*, nor does the brief sketch in the

of the Jesuits in Mauriac (Auvergne), where he was born in 1728 of noble, well-to-do parents, he eventually arrived in Paris to study at the Collège Louis-le-Grand. The director of the school recognized his talent for astronomy and mathematics and sent him to Cassini. Under Cassini's guidance he published one of the newly augmented editions of Halley's tables that began to appear in the mid-fifties of the century. Indeed, except for this, his observation of the comet of 1759, and his connection with the transits of Venus, he accomplished very little. But the intense efforts which he exerted on behalf of the transits, efforts that took him to a Siberian winter in 1761 and a southern California summer in 1769, also cost him his life, as we shall see; and the place he holds in history, his importance as a scientist, so to speak, therefore coincides with the decade of the transits. At that, his star glows briefly upon the eighteenth-century astronomical horizon more for the zeal and activity which it represents in the life of science than for the little light it sheds upon the problems of astronomy.

Accompanied by a colonel in the service of the King of Poland, Chappe d'Auteroche left Paris toward the end of November 1760, forced by the war to take an overland route to St. Petersburg, via Vienna and Warsaw. In view of the difficulty of moving "un appareil considérable d'Instrumens," it had been his original intention to go by sea, in spite of the war, by using a Dutch vessel, but late preparations had made him miss the last boat of the year, which was already at sea when he was ready to leave. The delay probably saved his life and certainly that of the expedition, for he later wrote: "J'eus bien-tôt lieu de m'en consoler, ayant appris, avant mon départ, qu'il [the Dutch boat] avoit échoué sur les côtes de Suede."[53]

Biographie Universelle, VII, 492-493, do any better. Even the biographical dossier at the Académie des sciences can add only odd bits, such as an extract from the baptismal registry at Mauriac and a signed receipt for the delivery of his pension by Buffon. Of greater value, though, are some letters pertinent to his death in 1769, about which I shall have more to say at another point.

[53] Chappe d'Auteroche, "Extrait du voyage fait en Sibérie . . . ," *Mémoires*, 1761, p. 337. This is a very brief account of his trip, read to the Académie 13 November 1762, after he had returned to Paris. A much fuller account was later published entitled: *Voyage en Sibérie, fait par ordre du Roi en 1761; contenant les moeurs, les usages des Russes, et l'état actuel de cette puissance;*

Heavy rains and a variety of accidents to their carriages made an eight-day trip of the distance from Paris to Strasbourg. All of the thermometers and barometers were broken when they arrived there, but while the carriages were being replaced, Chappe made a new set of instruments. Fear of a repetition of these events in Germany took him to Ulm where, in spite of advance warnings about the danger of navigating upon the river during its season of fog and flood, he embarked upon the Danube. Progress was slow, limited to daylight, and sometimes for only a few hours at that. But Chappe devoted his time to making a precise map of the river above Ulm, "knowing we had no particular map of the Danube in this part of its course."[54] At other times, when the boat was unable to proceed, he would climb the mountains that enclose the upper reaches of the river in order to determine their height with a barometer.

At Vienna, which he reached on 31 December 1760, Chappe was received by Joseph and Maria Theresa and visited the royal cabinet of natural history, where he was much impressed by the large coral collection. His entrance into the city itself was greatly facilitated by letters from the Hapsburg Ambassador at Paris to the director of customs. This was of great advantage in avoiding an inspection of his baggage that would have meant disturbing the careful packing of his instruments. Of even more importance, in Vienna he met the Jesuit astronomers, Hell and Liesganig, who were to observe the transit of Venus in that city. In addition to transit problems, they discussed the height of the mountains which Chappe had measured en route, compared and calibrated their barometers in a measurement of those around Vienna, and jointly determined the magnetic variation at Hell's observatory.[55]

When Chappe left Vienna on 8 January 1761, accompanied by Favier, the new Secretary of the French Embassy at St. Petersburg, the temperature averaged between 3 to 14 degrees below

---

la Description géographique & la nivellement de la route de Paris à Tobolsk (3 vols.; Paris: Debure, pere, 1768). This was translated into English by an unknown translator as A Journey Into Siberia. . . . (2nd edn.; London: Faden and Jeffreys, 1774).

[54] Chappe d'Auteroche, A Journey Into Siberia, p. 2.
[55] ibid., p. 7.

zero (Fahrenheit). Hills were icy and difficult to surmount, with frequent accidents and damaged equipment. Rivers not frozen solidly enough had to be crossed by first breaking a path through the ice for the ferry; frequently, when some were shallow enough, they were forded, with Chappe and Favier crossing on foot to break ice and lighten the load. By 19 January they were in Cracow, and by the 22nd in Warsaw. There Chappe learned from "le Resident Muscovite à la Cour de Pologne" that he was awaited in Russia with great impatience, and he so informed the Académie des sciences in a letter to Cassini de Thury.[56] They left Warsaw and crossed the frozen Vistula on 27 January 1761, and by 7 February, in Riga, they were forced to switch to sleds. "I experienced for the first time," Chappe wrote, "the ease of travelling with sledges; we went on with the greatest velocity, without meeting with any accidents."[57] At this new speed, they were in St. Petersburg within a week.

Upon arrival in St. Petersburg, Chappe immediately contacted the resident French minister, the Baron de Breteuil, and the Imperial Academy of Sciences. With the aid of Breteuil and the High Chancellor of Russia, Count Woronzof, he was able to obtain all the necessary cooperation for his trip across Russia, but he also learned from the Academy that his long delay in arriving had led them to believe that he would not make it to Tobolsk in time. Consequently, Russian observers had been sent out more than a month earlier, Paupof to Irkutsk near Lake Baikal and Rumovsky to Selenginsk.[58] All trace of the Paupof expedition seems to have disappeared, and no observation of the

---

[56] *Proc. Verb.*, 1761, fol. 33. The letter was read to the assembly on 11 February 1761.

[57] Chappe, *A Journey Into Siberia*, p. 25.

[58] Chappe, "Extrait du voyage fait en Sibérie . . . ," *op.cit.*, p. 338. Chappe actually said that Rumovsky was sent to Nertichinsk (Nerchinsk in modern spelling), which is farther east and closer to the Manchurian border, but Rumovsky's publications show that he never got there. For example, see S. Rumovsky, *Brevis exposito observationum occasione transitus Veneris per Solem in urbe Selenginsk* (Petropoli: Typis Academiae, 1762). This was also printed in the *Novi Commentarii*, XI (1765), 443-486. Curiously enough, in view of early Russian interest in the transits of Venus, they are not mentioned in the *Novi Commentarii* until 1764, X, 57-65. The city of Selenginsk, which I have been unable to locate under that name, is 7 hours, 6 minutes, and 25 seconds east of Greenwich, about 106.5 degrees, and about 51 degrees, 6 minutes north latitude.

transit of Venus was recorded at Irkutsk in 1761. But Rumovsky was successful, and we shall have occasion to refer to him again in discussing the results of 1761.

Because of the late date, some members of the Russian Academy suggested that Chappe choose another station, less distant and more accessible than Tobolsk, but since the duration of this transit would be shorter in this Siberian capital than anywhere else, he decided to attempt to reach his original destination. The Empress and the Court were most cooperative, and his departure for Tobolsk was set for 10 March 1761.[59] Difficult as the journey from Paris to St. Petersburg had been, the eight hundred leagues from St. Petersburg to Tobolsk were to be even more so, requiring preparations of another sort. Everything had to leave with the expedition: bread, beds, utensils, an interpreter, and even a watchmaker to repair anticipated damages to clocks and instruments. The one great fear was an early thaw, for only the speed attained by horse-drawn sleds over a smooth blanket of snow would make it possible to get to Tobolsk in time. Furthermore, a thaw would leave them stranded in the wilderness.

They left St. Petersburg on schedule, in four enclosed sleds, each drawn by five horses abreast.[60] Chappe and his personal provisions traveled in one, the watchmaker and Chappe's servant in another, a Russian sergeant sent by the government to serve as a guide in a third, and the instruments in the fourth.[61] In four days they were in Moscow, but the sleds were so damaged from the pounding they had received on the roads that they were beyond repair. A new set of sleds was obtained and, re-provisioned, they set out again on 17 March 1761. In the meantime, the Académie des sciences at Paris was kept informed of Chappe's progress by letters from the French ministry in Russia, the Secretary of the Imperial Academy, and Chappe himself. Thus in April and May they were thrice informed of what had taken place in Russia during February and March.[62] Upon one

[59] Chappe, A Journey Into Siberia, p. 26.
[60] ibid., p. 27.   [61] ibid., p. 28.
[62] Proc. Verb., 1761, fol. 73 (8 April 1761); fol. 91 (20 May 1761); and fol. 94 verso (23 May 1761).

such occasion, when La Caille presented to the Académie a letter which he had received from Muller (the Secretary of the Imperial Academy), announcing the arrival of Chappe at St. Petersburg and his subsequent departure for Moscow and Tobolsk, the Académie considered it timely to enter into the record of its deliberations, "que le Roi a bien voulu accorder à M. l'abbé Chappe 4000# pour son retour."[63]

The trip from Moscow to Tobolsk took about a month, toward the end of which it was a race between the oncoming spring with its accompanying thaw, and the speeding sleds. But fortunately the temperature stayed down, and at one part of the journey between the two important Volga River cities of Nizhni Novgorod (now known as Gorki) and Kozmodemiansk, Chappe tells us that "the surface of the Volga was as smooth as glass . . . and the sledges went on with inconceivable swiftness."[64] Leaving the Volga at the latter city and striking northeast meant, except for the Ural Mountain crossing near Solikamsk, a narrow road bordered by dense and lonely forests of birch and pine all the way to Tobolsk (some five hundred leagues). Only rarely was the isolation broken by a relief station for the horses and a hamlet of two or three houses. Chappe's narrative in this period is a description of the terrain, the life of the people (from their alleged licentiousness to the famous sweat-baths), odd commentaries on disease and the weather, and the adventure of finding himself deserted by his retinue in the midst of the Siberian forest and being forced to hunt them down pistol in hand.[65] With good fortune, he arrived in Tobolsk on 10 April 1761; the last river was crossed on ice that was already under water, and only six days before all of it broke up.[66]

With the help of the Governor of Tobolsk, who assigned a guard of three grenadiers and a sergeant to Chappe for the duration of his stay, the construction of an observatory atop a nearby mountain was immediately undertaken, and by 11 May 1761 it was ready for use. On 18 May he observed several phases of a lunar eclipse, and on 3 June he observed an eclipse of the sun that was invisible in France. These observations were ex-

[63] *ibid.*, fol. 91.   [64] Chappe, *A Journey Into Siberia*, p. 38.
[65] *ibid.*, pp. 46-47.   [66] *ibid.*, p. 77.

tremely important, for they enabled him to determine the longitude of his station, which, he tells us: "I could not expect to find from observing the eclipses of Jupiter's Satellites, because this hemisphere, in the summer time, is almost constantly lighted by the sun; and besides, this eclipse being visible also at Sweden, Denmark, and at St. Petersburg, I was sure of meeting with observations to answer mine."[67]

Chappe's work at the observatory during the day and a good part of the night, coming at a time when the river near Tobolsk was in flood, with heavy damage to the city and its inhabitants, brought him to the focus of local superstition and ignorance, which saw the event as a result of his mysterious doings. The guard which had been assigned him was increased, so that the rumblings of mob action never got to the point where either he or the observatory was seriously threatened. To satisfy the curiosity of the Governor, the Archbishop of Tobolsk, and the local nobility about a transit of Venus and to thank them for their cooperation, Chappe erected a tent near the observatory and placed a telescope in it for their use. This was also wisely conceived to leave him undisturbed for the important aspects of the observation. His excitement on the eve of the transit took him to the observatory to spend the night. "The sky was clear," he writes, "the sun sunk below the horizon, free from all vapors; the mild glimmering of the twilight, and the perfect stillness of the universe, completed my satisfaction and added to the serenity of my mind."[68] But Chappe could neither eat nor sleep that night, and every flaw in the heavens, every wisp of smoke or momentary cloud threatened his expressed serenity.

On 6 June 1761, Chappe assigned to the watchmaker who had accompanied him from St. Petersburg the task of recording the data and watching the clock, while the interpreter was to be employed in counting time. Excellent weather and the calmness of the day enabled him to remove the telescope from the building and observe in the open air. When he caught sight of Venus,

---

[67] *ibid.*, p. 78. Contact with Delisle gave him the necessary Swedish data, and he was able to calculate the longitude of Tobolsk as 4 hours, 23 minutes, and 34 seconds east of the Paris meridian.

[68] *ibid.*, pp. 80-81.

the planet was already partly immersed, so that he then prepared to observe the total entry, or internal contact: "The moment of the observation was now at hand; I was seized with an universal shivering, and was obliged to collect all my thoughts, in order not to miss it. At length I observed this phasis [sic], and felt an inward persuasion of the accuracy of my process. Pleasures of the like nature may sometimes be experienced; but at this instant, I truly enjoyed that of my observation, and was delighted with the hopes of its being still useful to posterity, when I had quitted this life."[69] In this respect, Chappe's observations more than fulfilled his wishes, for they were of great use to his contemporaries and continued to be employed to the end of the nineteenth century. During the three months in which he remained at Tobolsk after the transit, he continued to make observations of importance, although those of the transit of Venus were dispatched by mounted courier to St. Petersburg and Paris within a few days following the event. Some of those which he continued to make were directly related to the transit observation, such as the establishment of the precise latitude and longitude of Tobolsk, the parallelism of the fixed telescope on his quadrant, and the precise value of each turn of his micrometer screw. But at the same time he performed experiments on natural electricity, magnetic variations, and the length of the seconds pendulum;[70] and took advantage of his prolonged stay in the region to record various aspects of Siberian geology, meteorology, and natural history, including the social customs and mores of the peoples of the area.[71]

For his return, Chappe chose a more southerly route than the one used in going to Tobolsk. This took him through the Ural Mountain iron district, via the mining town of Ekaterinburg (known today as Sverdlovsk, and an iron ore center of gigantic proportions), where he visited the mines and commented at length on the variation and quality of the ores and the size of the workings.[72] His account of the return journey also contained

[69] *ibid.*, p. 83.
[70] Chappe, "Extrait du Voyage fait en Sibérie . . . ," *op.cit.*, 347.
[71] Chappe, *A Journey Into Siberia*, pp. 228-239, 297-319.
[72] *ibid.*, pp. 95-100, 164-204, 221-228. He described magnetite and haematite

descriptions of the lesser peoples of Russia whom he ⸱ncountered. And he was especially certain to point out the incredible backwardness of Russian society and the abject slavery of most of its population. Writing on the history of Russia for the period 1761-1762, which he spent in St. Petersburg following his return from Tobolsk, he gives an eyewitness account of the troubled times surrounding the change of regimes after the death of the Empress Elizabeth and before the accession of Peter III.[73]

Chappe's published impressions of his visit to Russia also included an essay on the progress of the arts and sciences there. Across the empire as a whole and even in such enlightened centers as St. Petersburg and Moscow, he found these to be at a very low state indeed. Men of ability invited from foreign parts, such as Bernoulli, Delisle, and Euler had gained a reputation for the Imperial Academy; but Chappe argued that in the several decades since Peter the Great and the policy of importing learned masters, "not one Russian has appeared . . . whose name deserves to be recorded in the history of the Arts and Sciences." And Chappe makes this statement while recognizing the contemporary genius of Lomonosow, "who would have made a considerable figure in any other Academy," and the potential in the young Stephan Rumovsky.[74] Even the foreign masters, Chappe felt, seem to have withered on the vine, and both the quality and quantity of their work had diminished in the bleak Russian atmosphere.

Chappe's explanation of this is a good example of the eighteenth-century scientific mind at work upon the raw material of history, the nature of man, and the institutions of an epoch. Two factors are predominant in Chappe's explanation of the Russian enigma, and these derive from Le Cat's physiology (the same Le Cat who was instrumental in creating the Académie des sciences at Rouen and in persuading Pingré to become an astronomer), and Montesquieu's theories about the impact of climate and geography upon the physical formation of a people.

and commented on the stratified layers in which they appeared. The workings in porphyry, gold, and marble of this region were also described.

[73] ibid., pp. 268-276.
[74] ibid., pp. 320-321.

Chappe is much concerned with what he calls the "nervous juice" that flows in man and is the source of all sensation as well as "the faculties of the soul." This physiological fluid, which Le Cat called the animal fluid, is in turn affected by a subtle fluid of the atmosphere which "all natural philosophers and anatomists place . . . in elementary fire; some call it the universal spirit, the vitriolic acid, the phlogiston, the electric matter &c."[75] Since the atmosphere is part of the climate, Chappe calls upon Montesquieu to support his contention, as indeed he does, that the climate of the North is such as to produce people having coarser organs, possessing fluids of a grosser kind than those of the South, and therefore more likely to produce men with large robust bodies rather than men of genius.

The shortcomings of the Russian empire in the eighteenth century were therefore explained in interrelated terms of physiology and environment. But not exclusively, for Chappe d'Auteroche would not be a child of his times if he did not also find grounds for the limited progress of the arts and sciences in Russia in the given structure of Russian society. "Why then is a Russian . . . so different from what he might be?" asks Chappe; and in the same breath replies, "The nature of education and of the government will furnish the solution of this problem."[76] Thus, while he proclaims the determinism inherent in the physiological and environmental theories of Le Cat and Montesquieu, Chappe, with the optimism of his century, sees a means of progress in the improvement of education and the destruction of despotic government, for "despotism debases the mind, damps the genius, and stifles every kind of sentiment."[77] It would not be germane to this study to elaborate further upon Chappe's analysis of Russia, save to point out that when published, it engendered much heat, including a line-by-line rebuttal by Catherine II.[78]

[75] *ibid.*, p. 322. Chappe spoke of the intimate connection of this substance with the life process, for he says of it, that "it is the primary fluid which gives life to the whole universe; but it is so subtle that it acts not upon our organs, but by the medium of the air. . . ."

[76] *ibid.*, p. 331.   [77] *ibid.*, p. 332.

[78] This was published anonymously and signed by A Lover of Truth. It was anonymously translated into English under the title, *The Antidote; or an Enquiry into the Merits of a Book, Entitled A Journey into Siberia, Made in MDCCLXI*

During the winter that he spent in St. Petersburg before returning to France, Chappe put together the final version of his memoir on the Tobolsk observations and had it printed there.[79] He read this version to the Imperial Academy in a meeting on 8 January 1762 and before leaving Russia, sent a copy to the Académie des sciences in Paris, which De Fouchy presented to the membership on 5 May 1762.[80] Back in Paris in November 1762, he read the *Extrait du voyage fait en Sibérie* to the Académie and then in January 1763 read them an addition to this memoir.[81] All of these reports contained essentially the same information, the numerical data which constitute the observations of the transit of Venus and their companion studies: that is, the tabular information from which the latitude and longitude of the station were determined, the pendular clocks regulated, the reading scale of the micrometers evaluated, and so forth.

Chappe's report also introduced a new consideration to the observations of the transit, for he records having seen a luminous ring around Venus. This naturally played an important part in determining the apparent diameter of the planet as well as the the moments of contact, though Chappe was not alone in making this observation. The Swedish observers at Stockholm and Cajaneborg noted it, as did Le Monnier at the Chateau de Saint-Hubert and De Fouchy in La Muette at the *Cabinet du Physique* of the King.[82] But we shall have more to say about this phe-

---

. . . *and Published* . . . *by the Abbé Chappe d'Auteroche*. . . . (London: S. Leacroft, 1772).

[79] Chappe d'Auteroche, *Mémoire du Passage de Venus sur le Soleil; contenant aussi quelques autres observations sur l'astronomie, et la déclinaison de la boussole* (St. Petersburg: Imprimerie de l'Académie imperiale des sciences, 1762).

[80] *Proc. Verb.*, 1762, fol. 189.

[81] Although they were read months apart, in 1762 and 1763, these two *mémoires*, the *Extrait du voyage fait en Sibérie* . . . and the *Addition au mémoire précédent, sur les remarques qui ont rapport à l'anneau lumineux, & sur le diamètre de Vénus, observé à Tobolsk le 6 Juin 1761* were both put in the *Mémoires* for 1761, which, however, were not printed until 1763.

[82] Chappe, "Extrait du voyage fait en Sibérie. . . ," *op.cit.*, p. 364. Le Monnier, "Observation du passage de Vénus sur le disque du Soleil faite au Chateau de Saint-Hubert en présence du Roi," *Mémoires*, 1761, pp. 72-76, and De Fouchy, "Observation du passage de Vénus sur le Soleil, faite à la Muette au cabinet de physique du Roi, le 6 Juin 1761," *ibid.*, pp. 96-105. Of the latter observation De Fouchy writes, "Je me rendis au cabinet de Physique . . . sous les ordres de M. le Marquis de Marigny, où j'avois été averti que Sa Majesté desiroit que l'observation fût faite. J'y trouvai M. Ferner, Professeur de l'Astronomie en

nomenon later. On the eve of his departure for Paris, every inducement was apparently offered him to remain at St. Petersburg, the Empress even offering him the same position that had once been held by Delisle, but Chappe refused.[83] In view of his opinions of Russia and the Russians published a few years later, he could not have honestly done otherwise.

III

The least successful of the three major French expeditions with respect to the original goal, the transit of Venus, was the one which was the first to leave and the last to return. Even before the debate on the transit of Venus had taken place at the meetings of the Académie des sciences, Le Gentil had left France, sailing from Brest on 26 March 1760 with the hope of establishing an observational station in the French colony of Pondichery.[84] By 10 July 1760 he was at the Isle de France, after a voyage that was rather uneventful, save for the loss of a fellow passenger by suicide and the pursuit by an English fleet near the Cape of Good Hope. They managed to escape the enemy, but were forced to go up the Mozambique Channel, around Madagascar, and come down upon the Isle de France from the north.[85] Whereas Pingré had used La Caille's method of lunar

---

Suède . . . qui s'y étoit rendu à même intention, & avec lequel je concertai tout ce que crumes nécessaires pour assurer le succès de l'opération nous convinmes de faire ensemble, & dont le résultat . . . lui appartient autant qu'à moi." Afterwards, Le Monnier made a brief addition to Chappe's observation of the diameter of Venus in "Considérations sur le diamètre de Vénus, observé à Tobolsk le 6 Juin 1761," *Mémoires*, 1761, pp. 332-333. See also Bengt Ferner, *Resa I Europa* (1758-1762), pp. 389-403; ed. Sten G. Lindberg for Lychnos-Bibliotek, Studier och Källskrifter Utgivna av Lärdomshistoriska Samfundet (Uppsala: Almqvist & Wiksells, 1956), Vol. 14.

[83] De Fouchy, "Éloge de M. l'Abbé Chappe," *op.cit.*, p. 168.

[84] See Chapter II for biographical details on Le Gentil and the background to his departure.

[85] Letter from Le Gentil dated Isle de France 15 September 1760, to De Lanux at Isle Bourbon, in Le Gentil, *Voyage dans les mers de l'Inde (1760-1771), à l'Occasion du Passage de Vénus sur le Disque du Soleil, le 6 Juin 1761 et le 3 du même mois 1769* (Paris: Imprimérie Royale, 1779-1781), II, 705. De Lanux was a resident of Isle de Bourbon who had been elected a correspondent of the Académie in 1754 and assigned to Réaumur. Both Pingré and Le Gentil kept up a heavy correspondence with him, Pingré's letters, by the way, being limited to two subjects: astronomy and freemasonry. De Lanux also corresponded with Buffon on the natural history of the island, and after Réaumur's death (in 1757), he was assigned to Bernard de Jussieu. See Lacroix, *op.cit.*, pp. 29-33.

distances to determine the longitude at sea during his voyage, Le Gentil used Pingré's technique of measuring the hourly angles of the moon. This apparently satisfied him, for in writing to the Académie after his arrival at the island, he said: "Je suis fondé à dire qu'en prenant toutes les precautions convenables et les positions de la Lune les plus avantageuses, on doit avoir et on aura (maintenant que la Théorie de la Lune est perféc-tionnee par M. Clairaut) la longitude Sur Mer à moins de dix lieuës près."[86]

Debarkation at Isle de France also gave Le Gentil a chance to learn of the miserable plight of French affairs in the Indian Ocean. Karikal (a French plantation on the Coromandel Coast below Pondichery), had recently been taken by the British, and Pondichery was besieged by land and blockaded by sea. A heavy French fleet which had been forming at Isle de France with the intention of raising the siege at Pondichery, by which means Le Gentil had also hoped to get there, had suffered irreparable damage in a sudden hurricane which struck the island on 27 January 1761. Most of the heavily armed vessels were severely damaged on the coral reefs, and there was much loss of life.[87] With the possibility of getting to Pondichery thus infinitely re-duced, Le Gentil informed the Académie in February 1761 that he was considering going to Batavia instead.[88]

But early in the month the contemplated voyage was canceled, though no reason was given. A letter to De Lanux, dated 6 February, suggests one, however, for he complained of a severe case of dysentery, and this would certainly have been reason enough for not making the trip. At the same time he considered going to Rodrigue, unaware, it seems, that Pingré was destined for that station. But he made no final decision, saying that "si d'ici à deux mois je ne trouve pas d'autre débouché, je suis résolu d'aller attaquer cette île [Rodrigue]. . . ."[89] Once again his plans were changed for him, this time by the arrival of orders from

[86] Letter to Grandjean de Fouchy, read at the Académie 14 February 1761, *Proc. Verb.*, 1761, fol. 36.

[87] *ibid.*, fol. 37.

[88] *ibid.*, fol. 38.

[89] Letter to De Lanux dated Isle de France, 6 February 1761, in Le Gentil, *Voyage dans les Mers de l'inde.* . . . II, 719.

France to send troops to Pondichery. He decided to go with the reenforcements, and so on 11 March 1761, he embarked for his original goal aboard the troopship *la Sylphide*. "J'avois donc," he wrote to De Lanux again in a spirit of optimism, "trois grands mois devant moi pour me rendre à la côte de Coromandel, & pour m'y préparer; tous les lieux m'étoient égaux; & il y en avoit beaucoup de neutres entres lesquels je pouvois choisir, en cas que Pondichéry fût bloqué par l'enemi."[90]

Thwarted by extremely bad weather, *la Sylphide* arrived off Mahe on the Malabar Coast on 20 May 1761, only to learn from a Moorish vessel encountered at sea that both Mahe and Pondichery had just been captured by the British. There was only one thing to do, and in all haste the ship turned around and retraced its route from the Isle de France.[91] It was thus from the bridge of *la Sylphide* on the high seas and under the clearest of blue skies that Le Gentil saw, but could not observe in the astronomical sense of the word, the transit of Venus of 6 June 1761. At Isle de France on 23 June, he learned that M. de Seligny, "Officier des Vaisseaux de la Compagnie des Indies," had observed the egress of Venus from the sun, and Le Gentil then determined the longitude of that station. He also discovered at that time that Pingré had gone to Rodrigue, and commented, "Je souhaite qu'il ait été plus heureux que moi."[92] Astronomically speaking, Pingré's fate was indeed a happier one, but even as Le Gentil was writing, Pingré was suffering, as we have already seen, another manifestation of British naval superiority in the Seven Years' War.

Cheated of the reward for all his enterprise, Le Gentil could not return to France empty-handed, and in a letter to the Duc de Chaulnes written shortly after the transit, he announced his intention of remaining in the area for another year in order to derive some use from his voyage at least for geography, navigation, and natural history.[93] That one year stretched into several,

[90] Letter to De Lanux dated Isle de France, 16 July 1761, *ibid.*, p. 721.
[91] *ibid.*, p. 738.
[92] *ibid.*, p. 760. The narrative of these events by Le Gentil is also to be found in the *précis-historique* which he published in the *Voyage dans les Mers de l'Inde*, II, 1-85.
[93] "Lettre à Mgr. le Duc de Chaulnes, pair de France," Arch. Acad. Sc.,

and before long Le Gentil had decided to stay until the next transit in 1769. In the eight years that intervened he made the Isle de France a kind of general headquarters for excursions to all parts of the Indian Ocean, from Madagascar to the East Indies, and even to the Philippine Islands.

The two huge volumes in which he published the results of this decade in the Indies are the fruits of a variety of labors. The first volume is primarily concerned with India, including a study of the customs and religions of its inhabitants. But more important from our present viewpoint, it deals with Brahmin astronomy as well. Although Le Gentil arrived at some rather strange ideas in this connection, such as the notion that Egyptian astronomy was only an imitation of Hindu astronomy, he did bring back much that was useful on the subject. It was just this kind of experience (Le Gentil made contact with Brahmin astronomers, who taught him their techniques) which helped others to draw up the picture of ancient and contemporary Eastern astronomy that the eighteenth century knew. Voltaire corresponded with him on the subject, and Jean-Sylvain Bailly thought highly enough of his research to make use of it in his own work.[94] Yet it would be false to leave the impression that all was approbation, for Delambre found fault with his theories on the origin of the zodiac and other still unanswered questions in the history of astronomy.[95]

In the second volume, Le Gentil described his work in Madagascar, the Mascarene, and the Philippine Islands. In all of his travels, he determined the precise latitude and longitude of important places, and studied the physical characteristics of the regions he visited. From the Philippines, for example, we have a study of the soil and its productions, an ethnography of the islands, a history of the Spanish colony there, and the exact geographic position of Manila. To these were added measure-

---

Dossier Le Gentil, fol. 1. The Académie des sciences was informed of this on 30 January 1762, the same day that it received Pingré's letter describing what had happened at Rodrigue, *Proc. Verb.*, 1762, fols. 18 recto-22.

[94] See J.-S. Bailly, *Histoire de l'astronomie ancienne depuis son origine jusqu'a l'établissement de l'école d'Alexandrie* (Paris: Chez les Freres Debure, 1775), pp. 96, 107, 109, 111-114, and 115.

[95] Delambre, *Histoire de l'astronomie au XVIIIe siècle*, pp. 702-706.

ments of the length of the seconds pendulum and variations of terrestrial magnetism in different places, as well as studies of winds, tides, monsoons, natural history, and the best navigational routes in the Indian Ocean. A rather high opinion of much of this work is still maintained, as witness the following by Alfred Lacroix:

D'une façon générale, on peut dire qu'en ce qui concerne la géographie, l'hydrographie, la physique du globe, Le Gentil a été un excellent observateur, précis et minutieux, joignant à ses constatations personnelles beaucoup de bon sens pour les interprétations et les comparaisons. A ce point de vue, il est très supérieur à la plupart de ses émules et contemporains; dans sa manière simple, sans grandiloquences, il montre pour les indigènes de tous les pays qu'il visita la plus intelligente sympathie.[96]

As for Le Gentil's experience with regard to the transit of 1769, it will be treated in connection with the general discussion of that event.

IV

Once under way, the overseas experiences of the English astronomers were in no manner comparable to those of the French, and we know very little of their actual voyages or the problems they faced in establishing their stations of observation. Mason and Dixon were thwarted in their ambition to reach Sumatra in the East Indies, first by their unfortunate adventure in the English Channel, and then by the French capture of Bencoolen, as we have already seen. Consequently, after their arrival at the Cape of Good Hope on 27 April 1761, they remained there to make their observations, rather than, like Le Gentil, seeking some equivalent station in the East Indies (Batavia for example), and thereby possibly facing the loss of all useful observations. Besides, only about six weeks remained before the transit was due, and the East Indies lay some six thousand miles away through war-torn waters and a hurricane season.

Immediately after their debarkation, Mason and Dixon set about constructing a temporary observatory and making the

[96] Lacroix, *op.cit.*, p. 23.

appropriate observations in preparation for the forthcoming transit.[97] They mounted their quadrant, began recording temperatures thrice daily—at morning, noon, and night—and set their astronomical clock going on 4 May 1761. Careful observations showed them that their clock was losing 2 minutes and 17 seconds per day. One technique which they used for checking their timepiece consisted of repeated observations of Antares in the constellation of the Scorpion and Altair in the Eagle. These bright stars were observed at equal altitudes above the east and west horizons, respectively; the highly consistent means resulting from these observations represented the times when the stars passed across the meridian of the station, and therefore could be used to control the reading of the clock.[98] On 18 May they observed an eclipse of the moon which, in addition to the usual observations of eclipses of Jupiter's satellites and occultations of stars by the moon, helped them to establish the exact longitude of their temporary station. In all of these activities and the observation of Venus itself, they carefully followed the instructions of the Astronomer Royal, James Bradley, briefly given as follows:

Locate the observatory where there is a clear view toward the northeast, north and northwest. Observe the first and second contacts of Venus with the limb of the Sun. Then measure the distance of Venus from the limb of the Sun to ascertain the nearest approach of Venus to the center of the sun's disk. Measure the diameter of Venus.

Set up the clock so that the observers at the telescopes are immediately accessible to it. Observers must be careful not to prejudice one another in their judgements of events and times. Make a preliminary trial of the clock with its pendulum adjusted as it was at Greenwich to ascertain how much it loses in a sidereal day. Then adjust it to solar time. Keep a record of the temperature in the clock case. Record how much the pendulum must be changed in length to keep solar time. . . .[99]

[97] Mason and Dixon, "Observations made at the Cape of Good Hope," *Phil. Trans.*, LII, 378-380.

[98] T. D. Cope, "The First Scientific Expedition of Charles Mason and Jeremiah Dixon," *op.cit.*, pp. 7-8.

[99] S. P. Rigaud (ed.), *Miscellaneous Works and Correspondence of the Rev. James Bradley* (Oxford: The University Press, 1832), pp. 388-390.

We have just noted how well Mason and Dixon followed these instructions with regard to the problem of time.

The day of the transit at Capetown was clear and serene. About two hours after sunrise, when Venus was approaching the western limb of the sun, visibility was excellent, and they independently recorded the times of internal and external contacts. They differed slightly; for Dixon internal contact had occurred 4 seconds earlier than it did for Mason, and for external contact, Dixon had seen it 2 seconds earlier. The latitude and longitude that Mason and Dixon established for their Capetown observatory were excellent, as indeed their transit observations had been, and years later, when the first issue of the *Nautical Almanac* came out in 1767 under the editorship of Maskelyne, he was able to write of their location of Capetown that "it is probable that the situation of few places is better determined."[100]

Mason and Dixon's observations from the Cape of Good Hope proved to be the only set of transit observations available for the South Atlantic region, for Maskelyne and Waddington fared rather badly at Saint Helena. Cloudy weather had prevailed at the island for about a month preceding the day of the transit, and they had been unable to make any useful observations. On 6 June 1761, they were only occasionally able to see Venus against the background of the sun, at which time they attempted to measure its position upon the disk and record the time, but their observations must be counted a failure.[101] Even the attempt to determine the parallax of Sirius met with no success, for the zenith-sector which had been hastily manufactured for the task proved to be defective.

But the ambitious program of astronomical research which Maskelyne had undertaken to pursue during the sixteen months in which he absented himself from England included other projects. These were more successfully completed. A tide gauge was set up in the harbor of Saint Helena, and a series of measurements of the rise and fall of the tide was made

[100] Quoted by Cope and Robinson, "Charles Mason, Jeremiah Dixon and the Royal Society," *op.cit.*, p. 57.
[101] Maskelyne, "Account of the Observations made on the Transit of Venus, June 6, 1761, in the island of St. Helena," *Phil. Trans.*, LII, 196ff.

between 12 November and 22 December 1761. By this time, Mason and Dixon had joined Maskelyne at the island, and Mason now assisted him in the tidal measurements. A calculation of the relative value of the gravitational constant was made, as it had been at the Cape of Good Hope, by means of the seconds pendulum and the results afterward compared with Greenwich. On the voyages to and from England, Maskelyne determined the longitude at sea by the method of lunar distances, with an error of about 1½ degrees in his result as compared to one of 7 degrees obtained by the ship's officers.[102]

The voyage which John Winthrop and his assistants made to Newfoundland was as successful as it was brief. At Saint John's they were given every assistance needed by the military commander of the garrison and Michael Gill, the "Chief Judge in the Courts." The town was bounded by high mountains, so that they were forced to search for a considerable length of time in order to find a place from which to observe the sun shortly after sunrise. Housing was provided by tents, and the clock and telescopes were mounted on heavy pillars set in the ground. The clock was regulated by taking corresponding altitudes of the sun. "[W]e repeated these operations," Winthrop wrote: "every fair day, and many times a day; and continued them with an assiduity which the infinite swarms of insects, that were in possession of the hill, were not able to abate, tho' they persecuted us severely and without intermission, both by day and by night, with their venomous stings."[103]

But in spite of this interference, they were well able to observe the egress of Venus from the sun on the day of the transit, and their observations have since carried considerable weight. In addition to the contact observation, Winthrop was able, in the interval permitted him, to determine five positions of the planet on the solar disk and therefore to lay off the path of its

[102] Cope and Robinson, "Charles Mason, Jeremiah Dixon and the Royal Society," op.cit., p. 58. Incidentally, the DNB article on Maskelyne gives him credit for introducing this method into navigation (Vol. 36, pp. 414-415), but we have already seen it in practice by La Caille during his voyage to the Cape of Good Hope (1750) and by Pingré in his voyage to Rodrigue.

[103] Winthrop, A Relation of a Voyage from Boston to Newfoundland . . . , p. 10.

transit across the sun, even though at Saint John's[104] part of it took place during the night of 5-6 June. He had some difficulty however, in collecting data to calculate the longitude. No eclipse of the sun or moon took place while he was on the island; Jupiter and its satellites were of no avail because they had not risen high enough to be well observed before dawn; and not until 11 June was he able to obtain a useful observation (an occultation of a star by the moon) for calculating the longitude of his observatory with respect to Greenwich.[105]

Thus we have examined the main overseas expeditions connected with the transit of 1761. Though most dramatic, these constitute only six teams of observers out of a very much larger number. Limited intracontinental expeditions were also dispatched, especially under the sponsorship of the Swedish Academy of Sciences and the Danish Crown. And in addition we have already noted the two independent Russian expeditions and the voyage of Cassini de Thury to Vienna. Since there were at least one hundred and twenty observers from sixty-two separate stations, any attempt to discuss them in a manner comparable to that given to the six British and French expeditions above would render this study inordinately long for the value gained. Moreover, there would result a definite loss of balance from an equal treatment of all observers and their stations, even if the documentary material for such a task were available, which it is not. The sense of urgency and enterprise in the eighteenth-century scientists connected with the transits was obviously focused on the long-range expeditions to extreme but important temporary stations. There was no special need to worry about established observatories taking part in so important an observation as that of a rare transit. Nor for that matter, was there anything extraordinary in the action of the national academies to find some means for sending astronomers to key points within their national boundaries or to nearby friendly powers. Consequently, the best way to reveal the extensive nature of these observations of the transit of Venus is to summarize the results in simple tabular form.

[104] *ibid.*, pp. 13-15.
[105] *ibid.*, p. 16.

The table which follows is based primarily upon the published memoirs on the transit, the work of J. F. Encke, who reconsidered the importance of the eighteenth-century transits of Venus in 1822, and of S. Newcomb, who did the same in 1890.[106] The particular form of the table and the selection of the material in it are my own. To avoid repetition of identical terms, the following abbreviations have been adopted. In the instruments column, "A" designates an achromatic telescope, "R" a a reflecting telescope; no designation, or one in feet, means that either the instrument used is unknown, or only the focal length is known of a nonachromatic telescope. By the "type of observation" is meant the two at ingress, outer and inner contact—respectively, I, 1 and I, 2; and the two at egress, inner and outer contact—respectively, E, 1 and E, 2. Whenever possible the source of the data will be given, such as, the *Novi Commentarii Academiae Scientiarum Imperialis Petropolitanae* (NCP), Hell's *Ephemerides meridianum Vindobonensem* (Eph. Vin.), and the German translations of the memoirs of the Swedish Academy, *Der Königlische Academie der Wissenschaften Abhandlungen* (Sch.Ab.). Other abbreviations used have already been given. Finally, the order of the table will follow the longitudinal position of the stations from east to west, except where it may be desirable to bring together the observations made in a single country.

OBSERVATIONS OF THE TRANSIT OF VENUS, 1761

PLACE	OBSERVERS	INSTRU-MENTS	NATIONAL SPONSOR/OR NATION-ALITY	TYPE OF OBSERVA-TION	SOURCE
1. Peking	Dollier	14'	French Jesuit	I,2;E,1-2	NCP, XI, 524
2. Selinginsk	Rumovsky	15'	Russian	E,1-2	Ibid., 455
3. Calcutta	Magee		British	I,2;E,1-2	Ph.Tr. 1761, 582

[106] J. F. Encke, *Die Entfernung der Sonne von der Erde aus dem Venusdurchgang von 1761 hergeleitet* (Gotha: Die Beckerschen Buchhandlung, 1822), and S. Newcomb, "Discussion of Observations of the Transits of Venus in 1761 and 1769," *United States Nautical Almanac, Astronomical Papers*, II, 5 (1890), 259-405.

PLACE	OBSERVERS	INSTRU- MENTS	NATIONAL SPONSOR/OR NATION- ALITY	TYPE OF OBSERVA- TION	SOURCE
4. Madras	Hirst	2'R	British	Ibid.	Ibid., 396
5. Ibid.	The Jesuits			Ibid.	NCP, XI, 569
6. Tranquebar	The Jesuits			Ibid.	Ibid.
7. Grand Mount	Duchoiselle		French	I,2;E,1	Encke, 88
8. Tobolsk	Chappe	19'	French	I,2;E,1-2	Mém. 1761, 361
9. Rodrigue	Pingré	18'	French	E,1	Ibid., 87, 443
10. Isle de France	De Seligny	8'	French	E,1-2	Le Gentil, Voy. aux Indes, II, 760
11. St. Petersburg	Braun	8'		I,2;E,1-2	Eph. Vind. 1762
12. Ibid.	Krasilnikow	6'	Russian	Ibid.	Ibid. & Sch.Ab. 1763, 138
13. Ibid.	Kurganow	2½'R	Russian	Ibid.	Ibid.
14. Cajaneborg	Planman	21'	Swedish	Ibid.	Sch.Ab. 1761, 156
15. Tornio	Hellant	20'	Swedish	Ibid.	Ibid., 181
16. Ibid.	Lagerbohm	32'	Swedish	Ibid.	Ibid.
17. Abo	Justander & Wallenius	20'	Swedish	Ibid.	Ibid., 158
18. Cape of Good Hope	Mason	2'R	British	E,1-2	Ph.Tr. 1761, 364
19. Ibid.	Dixon	2'R	British	Ibid.	Ibid.
20. Stockholm	Wargentin & Wilken	19'	Swedish	I,2;E,1-2	Sch.Ab.
21. Ibid.	Klingen- stierna	10'A	Swedish	Ibid.	Ibid.
22. Hernosand	Gissler	21'	Swedish	Ibid.	Ibid., 159
23. Ibid.	Ström	20'	Swedish	I,2;E,2	Ibid.
24. Uppsala	Mallet	1½'R	Swedish	I,2;E,1-2	Ibid., 143
25. Ibid.	Strömer	20'	Swedish	Ibid.	Ibid.
26. Ibid.	Melander	16'	Swedish	I,2;E,2	Ibid.
27. Ibid.	Bergmann	21'	Swedish	I,2;E,1-2	Ibid.
28. Calmar	Wickström	21'	Swedish	Ibid.	Ibid., 160
29. Carlscrona	Bergström	3'R	Swedish	E,1-2	Ibid., 161

136

PLACE	OBSERVERS	INSTRU-MENTS	NATIONAL SPONSOR/OR NATION-ALITY	TYPE OF OBSERVA-TION	SOURCE
30. Ibid.	Zegolström	21'	Swedish	Ibid.	Ibid.
31. Lund	Schenmark	21'	Swedish	E,2	Ibid., 163
32. Ibid.	Burmester	16"R	Swedish	E,2	Encke, 90-91
33. Landscrona	Brehmer	10'	Swedish	E,1-2	Sch.Ab. 163
34. Ibid.	Dehn	6'	Swedish	E,1	Ibid.
35. Ibid.	Landberg	21'	Swedish	E,1	Ibid.
36. Vienna	Hell	4½'R	German Jesuit	E,2	Eph. Vind. 1762
37. Ibid.	Herberth	12'	German	E,2	Ibid.
38. Ibid.	Rain	9'		E,2	Ibid.
39. Ibid.	Lysogorski	3'R		E,2	Ibid.
40. Ibid.	Cassini	9'	French	E,2	Ibid. & Mém. 1761, 409-412
41. Ibid.	Liesganig	11'	German Jesuit	E,1-2	Eph. Vind. 1762
42. Ibid.	Scherffer	4'R	German	E,2	Ibid.
43. Ibid.	Stein-Kellner	16'	German	E,2	Ibid.
44. Ibid.	Mastalier	13'		E,2	Ibid.
45. Ibid.	Müller	11'	German	E,2	Encke, 94-95
46. Tyrnau	Weiss	4'R	German	E,1-2	Eph. Vind. 1762
47. Laibach	Schöttl	16'	German	E,1-2	Ibid.
48. Wetzlas	von Schlug	4'R	German	E,1-2	Ibid.
49. Copenhagen	Horrebow	22'	Danish	E,1-2	Mém. 1761, 113-114
50. Trondheim	Bugge & Hascow	8'	Danish	E,1-2	Ibid.
51. Leipzig	Heinsius	Greg.R	Danish	E,1	NCP, X, 480
52. Munich		1½'R		E,1-2	Encke, 94-95
53. Ingolstadt	Kraz	16'	Jesuit	E,1-2	Ibid.
54. Ibid.		13'		E,1-2	Ibid.
55. Ibid.		11'		E,1-2	Ibid.
56. Rome	Audiffredi		Italian	E,1-2	Encke, 92

PLACE	OBSERVERS	INSTRU-MENTS	NATIONAL SPONSOR/OR NATION-ALITY	TYPE OF OBSERVA-TION	SOURCE
57. Bologna	Zanotti	2½′	Italian	E,1-2	Ph.Tr. 1761, 399; Eph. Vind. 1762
58. Ibid.	Marini	10′	Italian	E,1-2	Ibid.
59. Ibid.	Mateucci	22′	Italian	E,1-2	Ibid.
60. Ibid.	Frisi	6′	Italian	E,1-2	Ibid.
61. Ibid.	Cassali	8′	Italian	E,1-2	Ibid.
62. Ibid.	Canterzani	11′	Italian	E,1-2	Ibid.
63. Florence	Ximenes	4 1/3′R	Italian Jesuit	E,1-2	Eph. Vind. 1762
64. Dillingen	Hauser	18′	Jesuit	E,1-2	Ibid.
65. Göttingen	Mayer, T.	6′	German	E,1-2	Encke, 156-159
66. Würzburg	Hubert	1½′R		E,1-2	Eph. Vind 1762
67. Schwetzingen	Mayer, C.	10′A	German	E,1	Ph.Tr. 1764, 162
68. Nürnberg	Kordenbusch	Greg.R	German	E,1-2	Encke, 94
69. Klosterbergen	Silberschlag		German	E,1-2	Ibid.
70. Beyreuth	Grafenhahn		German	E,1-2	Ibid.
71. Regensburg				E,1-2	Ibid.
72. Leiden	Luloss	7′R	Dutch	E,1	Ph.Tr. 1761, 257
73. Lyon	Béraud	19′	French Jesuit(?)	E,1-2	Mém. 1761 473
74. Montpellier	Tandon	18′	French	E,1-2	Zach, Cor. Astr., 1 (1818), 24
75. Ibid.	Romieu	10′	French	E,1-2	Encke, 92
76. Ibid.	Roucher-Deratte	14′	French	E,1-2	Ibid.
77. Beziers	De Manse	3½′	French	E,1	Ibid.
78. Ibid.	Clauzade	7′	French	E,2	Ibid.
79. Chateau de St. Hubert	Le Monnier	18′	French	E,1-2	Mém. 1761 74-76
80. Ibid.	La Condamine	15″R	French	E,1-2	Ibid.
81. Paris, Obs.Roy.	Maraldi	15′	French	E,1-2	Ibid., 76
82. Ibid.	Belléri	6′	French	E,1-2	Ibid.

PLACE	OBSERVERS	INSTRU-MENTS	NATIONAL SPONSOR/OR NATION-ALITY	TYPE OF OBSERVA-TION	SOURCE
83. Ibid.	Zannoni	3½'	Italian	E,1-2	Ibid.
84. Ibid., H. de Clugny	Messier	R	French	E,1-2	Mém. Div. Savans, VI, 435
85. Ibid.	Libour	4½'R	French	E,1-2	Ibid.
86. Ibid.	Baudouin	25'	French	E,1-2	Ibid.
87. Ibid., Luxemb.	Lalande	18'	French	E,1-2	Mém. 1761, 83
88. Ibid., Col. Louis le Grand	Merville	6'R	French Jesuit	E,1-2	Ibid. 80-81
89. Ibid.	Clouet	2 2/3'R	French Jesuit	E,1-2	Ibid.
90. Ibid., Ecol.Mil.	Jeaurat	18'	French	E,2	Encke, 90-91
91. Ibid., Ste.Gen.	De Barros		Portuguese	E,1	Ibid., 92-93
92. Passy, La Muette	Ferner	28"R	Swedish	E,1-2	Mém. 1761, 99
93. Ibid.	Noël	4'R	French	E,1	Ibid.
94. Ibid.	De Fouchy	4'R	French	E,2	Ibid.
95. Conflans-s.-Carrière	La Caille	4½'A	French	E,1-2	Mém. 1761, 80
96. Ibid.	Bailly	6'	French	E,2	Ibid.
97. Ibid.	Turgot de Brucourt	12'	French	E,1-2	Ibid.
98. Vincennes	Prolange	19'		E,1-2	Encke, 92-93
99. Rouen	Bouin	16'	French	E,2	Mém. 1761, 43
100. Ibid.	Dulague	9'	French	E,2	Ibid.
101. Bayeux	Outhier	6'	French	E,1-2	Ibid., 133-134
102. Greenwich	Bliss	15'	British	E,1-2	Ph.Tr. 1761, 173
103. Ibid.	Green	2'R	British	E,1-2	Ibid.
104. Ibid.	Bird	1½'R	British	E,1-2	Ibid.
105. Hackney	Ellicot & Dolland	2'R	British	E,1-2	Eph. Vind. 1762
106. Clerkenwell	Heberden	2'R	British	E,1	Ibid.

	PLACE	OBSERVERS	INSTRU-MENTS	NATIONAL SPONSOR/OR NATION-ALITY	TYPE OF OBSERVA-TION	SOURCE
107.	London (Sav. House)	Short	2'R	British	E,1-2	Ph.Tr. 1761, 181-183
108.	Ibid.	Blair	1½'R	British	E,2	Ibid.
109.	Ibid., Spit.Sq.	Canton	1½'R	British	E,1-2	Ibid.
110.	Chelsea	Dunn	6'R	British	E,1-2	Ibid., 190
111.	Shirburn Castle	Hornsby	12'	British	E,1	Ibid., 176
112.	Ibid.	Phelps	14'	British	E,1	Ibid.
113.	Liskeard	Haydon		British	E,1-2	Ibid., 203
114.	Madrid	Rieger	8'	German Jesuit	E,1-2	Eph. Vind 1762
115.	Ibid.	Eximenus	2 1/2' Quad.		E,1-2	Newcomb, *op.cit.*, 294-295
116.	Ibid.	Benevent	2¾'R		E,1-2	Encke, 92-93
117.	Porto	De Almeida	2'R	Portuguese	E,1-2	Mém. Div. Savans, VI, 352
118.	Lisbon	Ciera		Portuguese	E,1-2	Encke, 92-93
119.	Constanti-nople	Porter	1½'R	British	E,1-2	Ibid., 94-95
120.	St. John's, N.F.	Winthrop		British	E,1-2	Ph.Tr., 1764, 279

The above tabulation does not, in any manner, pretend to represent all the observations that were made of the 1761 transit. Others were known to have been made in various locations; even the name of the observer is sometimes known in these cases, but somehow the results never found their way into the standard literature of contemporary astronomy. Encke, for example, mentions thirty-nine such stations scattered throughout the world. It would be possible at this writing, to reduce that list by citing the appropriate data available from some of the research behind this study, such as William Chapple's observations at Exeter,[107] and Cardinal de Luynes's at Sens.[108] Or, it would be equally possible to add the names and places of other observations that were completely unknown to Encke, such as Niccolò Carcani's at Naples,[109] and Dirk Klinkenberg's at The Hague.[110] But in speaking of these unknowns, Encke made the point that "der gröste Theil . . . sind englische und französische Provinzialstädte, in welchen . . . kein Astronom von Ruf beobachet hat," and at the same time that "von den deutschen dürfte keiner dieser Orte einer genauen Beobachtung sich rühmen können."[111] The introduction of these additional observations at this time would simply be extraneous, since they played no part in the final calculations of the solar parallax. Indeed, the data summarized in the above table (implying the actual numerical data of the observations) represents the supply available to those who set out to calculate the solar parallax after the transit.[112]

Without overemphasizing its meaning, the statistical breakdown of this table is rather revealing. As might be expected, the French led all the others in the number of observers with thirty-one, but at the same time this is more than quantitative, for the occurrence of names like Le Monnier, Messier, Lalande, and La Caille indicates the participation of astronomers of the first rank. Surprisingly enough the Swedes, with twenty-one observers, displace the British from the second position which one would have expected them to occupy, for the British could muster only nineteen successful observers. Here, too, the displacement seems to be one of quality. Though we know little of most of the Swedes involved, the majority of the British observers, although

107 W. Chapple, "Observation of the Transit at Exeter," *Gentleman's Magazine*, xxxi (June 1761), 248, and (August 1761), 357-359.
108 Cardinal de Luynes read his transit observations to the members of the Académie on 20 June 1761. *Proc. Verb.*, 1761, fol. 117.
109 Niccolò Maria Carcani, *Passagio di Venere sotto il Sole, osservato in Napoli nel Real Collegio delle Scuole Pie, la mattina de' 6 Giugno 1761* (No place, publisher, or date).
110 Letter from D. Klinkenberg to Delisle, dated 17 June 1761, Dép. Marine, Vol. 115 (xvi-8), No. 5a-b, fols. 1-5.
111 Encke, *op.cit.*, p. 19.
112 These will simply be included in the final bibliography. Of course there were many attempts to observe the transit that failed, such as those of Le Gentil and Boscovich, but they too have no place in the table.

men of outstanding ability, were nevertheless amateurs. Incidentally, it is worth noting that the British list includes four of the leading British instrument makers, Bird, Ellicott, Dolland and Short. It is a kind of measure of the high level of their scientific interests that they should so take part in the transit observations. As a matter of fact, this sort of activity is more typical of British instrument makers than any other.

If the general run of Swedish observers were comparable in caliber to Pehr Wilhelm Wargentin (1717-1783) or Torbern Olaf Bergmann (1735-1784), then they easily deserve to be ranked in the upper levels with the French. Wargentin, who was Secretary of the Swedish Academy of Sciences for thirty-four years, was the organizing genius behind the many excellently dispersed stations to which the Swedish Academy, with royal funds, sent its observers. Indeed, as Linnaeus said, "the whole blossoming and existence of the Academy of Science" itself rested upon the work of Wargentin.[113] Wargentin's main astronomical interest was in the satellites of Jupiter, though he faithfully corresponded with Delisle on a variety of astronomical problems, as we have already noted in Chapter II. As a matter of fact, he had once intended (toward the end of 1745) to go to St. Petersburg to study astronomy under Delisle, but canceled his plans when he learned that Delisle was returning to France.[114] His astronomical training was nevertheless well-grounded, for he received it from the hands of the eminent Anders Celsius. Torbern Bergmann was a lecturer in physics at Uppsala when he was called to take part in the transit observations. Besides astronomy, his writings include studies on the rainbow, the northern lights, and pioneer work in physical geography. His great reputation, however, was made in chemistry, especially as a result of his theory of chemical affinity, which was held for a considerable time until it was replaced by the law of mass action.[115] With all this, he was also a very competent astronomical observer.

[113] S. Lindroth, "Pehr Wilhelm Wargentin," in S. Lindroth (ed.), *Swedish Men of Science, 1650-1950* (Stockholm: Almqvist and Wiksells, 1952), p. 108.
[114] N. V. E. Nordenmark, *Pehr Wilhelm Wargentin, Kungl. Vetenskapsakademiens Sekreterare Och Astronom 1749-1783* (Uppsala: Almqvist & Wiksells, 1939), pp. 312-313.
[115] Hugo Olsson, "Torbern Bergmann," *Swedish Men of Science 1650-1950*, pp. 131-138.

Because of their northern position the Swedish stations were of great importance; it was their observations which confirmed two unexpected phenomena in the transits—a luminous ring around Venus, and what has since been called the "black-drop" effect—and helped to upset their ultimate, definitive value. These phenomena were seen by Chappe, Le Monnier, De Fouchy, and others almost everywhere. For example, Mallet's report, which came from Uppsala, followed up the description of these phenomena with an explanation:

De ce qui est décrit, nous avons jugé que Vénus est vraisemblement pourvue d'une atmosphère qui, par suite d'une réfraction assez vive des rayons du Soleil, a rendu ces phenomènes inattendus si remarquable: et il a été avancé qu'aucun autre moment ne devait être considéré comme étant celui du contact intérieur de Vénus avec le bord du Soleil que celui où la "bande noire" semblait se rompre, quand Vénus se détachait du bord du soleil. . . .[116]

But let us momentarily postpone discussion of this problem of determining contact, so crucial to any observation of the transit, and return to the statistical conclusions of the table.

Following the one-two-three position of the French, Swedish, and British observers, the fourth place is occupied by the German-speaking group. Most of these were Jesuits, who made observations in the regions around the schools to which they were attached, usually as professors of astronomy and mathematics. As a whole, they were fairly well equipped and their observations were reliable. Indeed their ranks include two distinguished names, Maximilian Hell and the great Tobias Mayer (who was not a Jesuit), whose lunar tables were the best of the century.[117] Hell observed in Vienna, where he had been appointed director of the University's observatory in 1756, and where for thirty-six years the famous *Ephemerides Vindobonensis* were turned out under his editorship.[118]

The Italians make up the next largest group with nine observers, followed by the Russians, Danes, and Portuguese with three each. An unclassified group of sixteen remains. Of the

[116] Quoted in the French summary of Nordenmark, *op.cit.*, p. 322.
[117] Mayer's printed observations exist only in Encke's book, communicated to him by Baron von Zach, who copied them from Mayer's *tagebuch*. Encke, *op.cit.*, pp. 17, 156-159.
[118] G. Sarton, "Vindication of Father Hell," *Isis*, xxxv, 97.

Italians, Leonardo Ximenes (1716-1786) and Eustachio Zanotti (1709-1782) are easily the most outstanding, with G. B. Audiffredi (1714-1794) a close third. A Jesuit, like the majority of his fellow countrymen in science, Ximenes spent most of his life in Florence, where he alternately served the University and the Grand Duke of Tuscany. In the former capacity, he taught and wrote on astronomy, geometry, hydraulics, and mechanics. He was also responsible for the construction of the observatory of San Giovannino de Firenze and the famous gnomen of Florence.[119] Zanotti was an astronomer long connected with the Bologna Institute, eventually becoming its president. He was in correspondence with a number of leading European scientists, including La Caille, who called upon him to verify his Cape of Good Hope calculations. He was also a foreign correspondent of the Berlin Academy and of the Royal Society.[120] Audiffredi's importance is based more on his calculation of the solar parallax than his actual observation of the transit, and we shall have occasion to refer to that later.[121]

Not too much can be said of the remaining observations, those of the Russians, Danes, and Portuguese and the various observations of unknown nationalities. The astronomers engaged seem to have been adequate for the task, and their memoirs are straightforward compilations of data without special color or interest, though Christian Horrebow did publish a separate article on the path of Venus across the sun.[122] Of this entire group, the observation of the Russian, Stephan Rumovsky, was most important because of the extreme eastern position of his sta-

---

[119] "Leonardo Ximenes," *Enciclopedia Italiana*, xxv (1937), 824. Because of his Spanish birth (Ximenes is really a Spanish name), there is also an article on him in the *Enciclopedia Universal Illustrada, Europeo-Americana*, LXX (1930), 567. There is also an *éloge* on him by Luigi Caccianemici Palcani in the *Atti dell' Instituto delle Scienze di Bologna* for 1791, but I have not had access to it. Finally, there is a complete bibliography, numbering fifty-one items, mostly on hydraulics, in C. Somervogel, *Bibliothèque de la Compagnie de Jésus* (Paris: Picard, 1898), pp. 1341-1351.

[120] "Eustachio Zanotti," *Nouvelle Biographie Générale*, XLVI, 954.

[121] Encke, *op.cit.*, p. 32. He published two pamphlets on this problem, *Investigatio Parallaxis Solaris* (Rome: 1765), which, strangely enough, he signed with the acrostic name, Dadei Ruffi, and *De Solis Parallaxi* (Rome: 1766).

[122] C. Horrebow, *Dissertatio de semita, quam in Sole descripsit Venus per eundem transeundo die 6 Junii Ao. 1761* (Havniae: Typis Directoris S.R.M. & Universitatis Typogr. Nicolai Christiani Höpffneri, 1761).

tion. This is true even in the face of the more easterly station of the Jesuits in Peking because the Peking observations turned out to be of little value. Rumovsky was also the only member of this latter group to undertake the calculation of the solar parallax.

What then were the conclusions to be drawn from the 1761 transit? The first attempt to answer that question in the form of a specific evaluation of the solar parallax is given to us as a *fait accompli* in the extract of a letter from Lalande to Maskelyne, which was printed in the *Philosophical Transactions* of the Royal Society. Offhand we are immediately given two values and the implications that there had been earlier correspondence between them on this matter. For Lalande writes on 18 November 1762 that "M. Pingré, who is returned from the island of Rodrigue, has found the parallax of the sun to be the same as I have done; namely 9″ 2/5. I am not surprised that you find it to be only 8″ 3/5, since the Swedish observations, which appear to me to be very good, make it still less than you have found it. These uncertainties arise from our not having the difference of the meridians of the Cape, Rodrigues, Tobolski, Paris and London well determined."[123] The figure which Lalande quoted to Maskelyne is actually different from those which he and Pingré had separately calculated, or the one which De Fouchy put into the *Histoire* as the average of all the results to date. Thus Lalande's value was actually 9.55 seconds,[124] Pingré's 10.60 seconds,[125] and De Fouchy's 9.16 seconds.[126]

These conclusions were very rapidly followed by an extensive treatment of the solar parallax in two papers by James Short,[127]

[123] Lalande, "Extract of a Letter from M. de la Lande, at Paris, to the Rev. Nevil Maskelyne, F.R.S. Dated Paris, Nov. 18, 1762," *Phil. Trans. Abgd.*, xi, 648.

[124] Lalande, "Remarques sur les observations faites par M. Pingré à l'isle Rodrigue dans l'océan Ethiopique, pour la parallaxe du Soleil," *Mémoires*, 1761, p. 95.

[125] Pingré, "Observations astronomiques pour la détermination de la parallaxe du Soleil, faites en l'Isle Rodrigue," *ibid.*, p. 486.

[126] De Fouchy, "Sur la conjonction écliptique de Vénus et du Soleil du 6 Juin 1761," *Histoire*, 1761, 116.

[127] J. Short, "The Observations of the Internal Contact of Venus with the Sun's Limb, in the late Transit, made in different places of Europe, compared with the Time of the same Contact observed at the Cape of Good Hope, and the Parallax of the Sun from thence determined," *Phil. Trans. Abgd.*, xi, 649-660,

in the *Philosophical Transactions*. To calculate the parallax, Short used several involved techniques. In the first of these, he chose certain key stations and then made selected comparisons with them. Thus he made 18 comparisons with Cajaneborg, 17 with Bologna, and 18 with Tobolsk, for a total of 53 comparisons to obtain a parallax of 8.58 seconds. Then he repeated this procedure 63 times with other stations to get a parallax of 8.55 seconds. He averaged these two results from a total of 116 comparisons[128] to get a mean of 8.565 seconds for the solar parallax. A second method consisted of averaging the mean of 21 comparisons of internal contact with the results of the observation at the Cape of Good Hope to get a mean parallax of 8.56 seconds. For a third method, he repeated the second, using Rodrigue, to get a mean parallax of 8.57 seconds. In the fourth method, an average of the comparisons of the observed total durations yielded a mean parallax of 8.61 seconds. Two additional averages were arrived at from comparative considerations of the apparent least distances of the centers of the planet and the sun. These mean parallaxes were 8.56 and 8.53 seconds. A mean of these means was then taken to give a concluding value of 8.56 seconds to the solar parallax. All the results of his varied methods had so nearly coincided that it seemed to Short to be "impossible that the mean of them all can err 1/10 of a second, and that probably the error does not exceed 1/500 part of the whole quantity, as Dr. Halley had many years since confidently presaged."[129]

Had these results been universally accepted, no further discussion would be necessary here, but Short found no scientific consensus anywhere, and he alone upheld the prescient confidence of Halley. Between the publication of Short's two papers, Pingré published his own conclusion about the solar parallax, one which differed from Short's by at least two seconds, as we have just seen. This precipitated a rather strong debate between the two, which for a while took on nationalistic overtones.[130]

---

and "Second Paper concerning the Parallax of the Sun determined from the Observations of the late Transit of Venus; in which the Subject is treated at Length, and the Quantity of the Parallax more fully ascertained," *ibid.*, XII, 22-37.
[128] Short, "Second Paper concerning the Parallax of the Sun. . . ," *op.cit.*, p. 25.
[129] *ibid.*, p. 34.   [130] See Encke, *op.cit.*, p. 30.

One of the factors that had rendered Short's results so homogeneous had been the rather judicious series of alterations which he had made in the original data concerning longitude and time of contact at various stations. Later on, when Pingré replied to Short's paper of 1763, he wrote of this action: "Je crois avoir assez détruit le concert prétendu des combinaisons de l'auteur de Mémoire, puisque j'ai montré que la plupart de ces combinaisons étoient appuyées sur des observations incertaines & mêmes alterées. . . ."[131]

But this debate on the value of the solar parallax to be determined from the 1761 transit was not limited to Pingré and Short alone. Almost immediately after the issue of Short's papers, Thomas Hornsby, Savillian Professor of Astronomy at Oxford, published a value of his own which was closer to the French conclusions than to those of his fellow countryman. Hornsby decided that the solar parallax was 9.732 seconds, adding the supporting argument that "in this quantity of the sun's parallax we must either acquiesce, or remain as ignorant of the true quantity of it as we were before, till we can have recourse to the next transit on 3 June 1769. . . ."[132]

But Hornsby's demand for submission to his figure failed to bring new calculations to a halt. And to those already cited, additional evaluations were added from sources as varied in origin as they were divergent in their conclusions. Thus, Stephan Rumovsky arrived at a solar parallax of 8.33 seconds,[133] whereas Anders Planman of the Swedish Academy of Sciences said that "the sun's parallax may be stated at 8" 28, so far as the difference of meridians can be depended on." But he also thought this value would prove to be too large after the next transit.[134] Fi-

---

[131] Pingré, "Nouvelle recherche sur la détermination de la parallaxe du Soleil par le passage de Vénus du 6 Juin 1761," *Mémoires*, 1765, p. 23.

[132] T. Hornsby, "On the Parallax of the Sun," *Phil. Trans. Abgd.*, XII, 60.

[133] S. Rumovsky, "Investigatio parallaxeos solis ex observatione transitus Veneris per discum Solis Selenginski habita, collata cum observationibus alibi institutis," *Novi Commentarii*, XI (1765), 487-538. Also, "Animadversiones in supplementum cal. *Pingré* ad dissertationem eius de parallaxi Solis," *ibid.*, XII (1766-1767), 575-586.

[134] A. Planman, "A Determination of the Solar Parallax attempted by a peculiar Method, from the Observations of the last Transit of Venus," *Phil. Trans. Abgd.*, XII, 527. This memoir was first printed in the Swedish Transactions for 1763 and 1764; see Encke, *op.cit.*, p. 31.

nally, the Roman astronomer, G. B. Audiffredi, published two papers on the solar parallax, concluding that its value was 9.25 seconds.[135]

Consequently, in spite of all the care which had gone into the preparations for the observations of the transit of Venus, and the tremendous world-wide effort to render them successful, the dream of a definitive determination of the solar parallax seemed as remote as it had always been. Figures that ran from 8.28 to 10.60 seconds were no measure of certainty whatever. One difficulty upon which these transit observations foundered, as had already been suggested, centered upon the determination of the precise moment of contact between Venus and the sun. The apparent entry of Venus upon the surface of the sun was seen clearly enough, but when it came to determining the exact instant of internal contact, the observers were upset to discover that the planet appeared to lose its perfect circular form. Diagram 3 will help to illustrate this point.

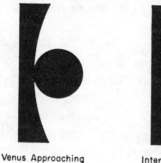  

Venus Approaching   Internal Contact   The Black Drop
Internal Contact   or Ligament

3. VENUS AND THE SUN AT INTERNAL CONTACT

In the first drawing, Venus is already well upon the sun's disk, moving from left to right. The moment of internal contact, defined as the instant at which a continuous thread of light appears between the planet and the sun's limb, is ideally pictured in the second drawing. This very desirable image was never seen. An approximation to what the observers actually saw appears in the third drawing. The matter of the planet ap-

[135] See note 121.

peared to stick to the sun's edge like taffy candy, only gradually pulling apart as the planet moved across the sun. Moreover, the effect varied from one observer to the next, making it virtually impossible to determine the moment of contact precisely. This effect, and the reasons behind it, will again be discussed in connection with the transit of 1769, though obviously one explanation lies in the effect of the Venusian atmosphere in dispersing the rays of light coming from the sun.

The problem of determining the exact longitude of the many stations was the second major factor in bringing the 1761 enterprise to grief. Much of the argument between particular evaluations of the solar parallax is to be found in differences of opinion about the position of a station. This was no small problem in the eighteenth century, and indeed nothing can better illustrate this than another fragment from Lalande's letter to Maskelyne about the solar parallax. The immediate problem under discussion is the longitudinal difference between Greenwich and Paris, to be determined by the occultations of various stars by Mars and the moon. From these, Lalande writes, "You may deduce the difference of the meridians of these two cities, which we may be ashamed to say we are uncertain of to 20 seconds."[136] Here then, in 1762, not even the longitudinal difference between these two major observational centers was well established. Nevertheless, in the face of all these difficulties, a spirit of optimism prevailed about the success that would certainly come with the transit of 1769. Indeed, that optimism was more than spiritual, for the actual undertaking in the name of the second transit of the decade was even greater than the first.

[136] Lalande, "Extract of a letter from M. de la Lande . . . to Nevil Maskelyne . . . ," *op.cit.*, p. 649.

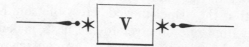

THE TRANSIT OF 1769 AND

FINAL CONCLUSIONS

INTEREST in the transits of Venus as a whole did not slacken after the transit of 6 June 1761 had become history. If anything, the contrary effect took place: the waves of popular interest produced by dropping the question of the solar distance into the pool of unsolved scientific problems in the middle of the eighteenth century continued to reach out on an ever-widening front. Various circumstances favored this development. The debate on the specific value of the solar parallax deducible from the observations of the 1761 transit persisted throughout the decade without definitive conclusions, so that the issue was kept open and very much alive. The unique feature of having rare and important astronomical events occur in pairs within a relatively short interval of time did of itself guarantee a sustained interest in the transits. Furthermore, the combination of the inconclusive results of 1761 and the proximity in time of the transit of 1769 as a kind of second chance to solve the problem demanded a renewed scientific activity of even greater dimensions. And finally, as a promise of success, and therefore a further stimulant to exploit the second transit in spite of the failure of the first, there was the knowledge that the 1769 transit would be easier to study astronomically because of its more favorable position on the solar disk.

French concern with the transit of 1769 remained as high as it had been for that of 1761. Indeed, in planning for the first transit Lalande had already begun the discussion of the second by preparing a world map showing the effect of parallax on the times of ingress and egress at different places on the globe in 1769 (Figure 10).[1] This paper was read to the Académie in

[1] Lalande, "Mémoire sur les passages de Vénus . . . en 1761 et 1769 . . . ," *Mémoires*, 1757, pp. 232-250.

May 1760, well before both transits and the experience acquired in 1761. But several years later, in March 1764, he presented the Académie with a new map and memoir on the same subject, which Bailly and De Fouchy were charged with examining.[2] This presentation took place during the very period when Pingré and others were discussing the results of 1761, and it is typical of what was happening during the interval between the two transits. The results of the first transit and the prospects of the second were often simultaneously discussed.

While these activities were under way in Paris, Le Gentil, already on location, so to speak, was again preparing to observe the transit. Back at Isle de France after trips to Madagascar and some of the islands of the Mascarenes, collecting the kind of information that went into the narrative of his stay in the Indies which we have already discussed, Le Gentil informs us that "ces differentes occupations m'avoient conduit jusqu'a l'année 1765. Il étoit temps de penser au second passage de Vénus."[3] These thoughts included a recalculation of the transit for observations from India, the Philippine Islands, the Marianas, Mexico, and even Europe. From this operation, Le Gentil concluded that a station in Manila or the Marianas would be the best of all choices open to him, not because the effect of parallax there would be any greater than at Pondichery or the Coromandel Coast of India, "mais parce que l'élévation du Soleil sur l'horizon au moment de la sortie de Vénus, devoit être fort grande, & donnoit par ce moyen plus d'espérance de réussir, qu'on ne pouvoit espérer de faire à la côte de Coromandel, où le Soleil devoit être fort bas au moment de la sortie de Vénus. . . ."[4]

Accordingly, Le Gentil wrote to Clairaut, Lalande, and the Duc de Chaulnes in January 1766 of his intentions, asking the

[2] *Proc. Verb.*, 1764, fol. 56. By May, it had been approved and printed, and Lalande presented it to the Académie, *ibid.*, fol. 205. It was published as *Figure du passage de Vénus sur le disque du Soleil, qui s'observera le 3 Juin 1769* (Paris: 1764), and again in Amsterdam with a slightly different title in 1769. After the transit, Lalande published a sort of anthology of the actual observations to go with this earlier publication: *Mémoire sur le passage de Vénus, observé le 3 Juin 1769; pour servir de suite à l'explication de la carte publiée en 1764* (Paris: n.p., 1772).

[3] Le Gentil, *Voyage dans les Mers de l'Inde* . . . , p. 16.

[4] *ibid.*, p. 17.

latter to obtain letters for him from the Spanish Court to its Governor in the Philippines. By June of the same year, the letters were read to the members of the Académie, but no comment either approving or disapproving the project was forthcoming at that time.[5] But even before the Académie had received his letters, Le Gentil was on his way to the Philippines, having taken advantage of the opportunity offered by the rare presence at Isle de France of a Spanish man-of-war bound for Manila from Cadiz. In the Philippine capital by 10 August 1766, Le Gentil still hoped to get to the Marianas, but upon learning that connection with the islands was made only at about three-year intervals, he decided to remain in Manila.[6] As he had done elsewhere, he determined the longitude of the city by means of immersions of the satellites of Jupiter, and later published a separate memoir on the subject which included a study of the length of the seconds pendulum at the same location.[7]

At Paris, in the meantime, Pingré drew up a memoir on the choice of stations for the coming transit which he read to the Académie in three installments between December 1766 and February 1767. Partly based on Lalande's earlier work, especially his map, it also took cognizance of similar English writings, and included a special geographical discussion that dealt at length with the desirability of having observations made from the South Seas.[8] By reviewing the history of the earlier voyages of exploration in these waters by men like Mendana, Juan Fernandez de Quiros, and even Magellan, Pingré argued that there was every likelihood that land for an observational station could be found there, for all of these men had spoken of various groups of

[5] *Proc. Verb.*, 1766, fol. 199 (11 June), and fol. 212 (25 June).

[6] Le Gentil, *Voyage dans les Mers de l'Inde . . .* , p. 19.

[7] Le Gentil, "Observations astronomiques faites pour déterminer la longitude de Manille," *Mémoires*, 1768, pp. 237-246.

[8] Pingré, *Mémoire sur le choix et l'état des lieux où le passage de Vénus du 3 Juin 1769 pourra être observé avec le plus d'avantage; et principalement sur la position géographique des isles de la mer du sud* (Paris: P. G. Cavelier, 1767). This separate printing was authorized by the Académie, after it was carefully examined by Chappe and Lalande acting as an official committee, *Proc. Verb.*, 1767, fols. 48-50 verso. Discussion of it also took place in the *Histoire* for 1767, pp. 105-109, and it was favorably reviewed in the *Journal de Trévoux* (Juillet 1767), pp. 145-147.

islands in the region, though little beyond that was known about them. He even suggested the possibility of using the strange and remote Easter Island. And to support this argument as a whole, Pingré drew a map of the South Seas and attached it to his memoir (Figure 14).

The major reason for stressing a South Seas observation for 1769 was a direct result of the 1761 experience. Having examined all the methods for determining the solar parallax in the bitter debate with Short, Pingré came to believe—and virtually all contemporary astronomical opinion concurred in that belief —that the method of durations would best serve the calculation of a definitive solar parallax. As he forcefully expressed it in a communication after the second transit: "La méthode la plus simple, la plus naturelle, la plus facile, la moins exposé aux contradictions, de toutes celles que l'on peut employer, pour conclure la parallaxe du Soleil d'un passage de Vénus sur le disque de cet Astre, est celle où l'on ne fait usage que de la différence des durées observées en des lieux où l'effet de la parallaxe ait le plus sensiblement différé."[9] It was therefore the search for a station where the total duration of the transit would be shortest that drew Pingré to focus upon the South Seas, for it was already known that the longest duration would be observed from Lapland, probably at a station like Torneo. For a theoretical station at about 242 degrees east longitude and 28.5 degrees south latitude, using Pingré's own map, he estimated that there would be a duration of 5 hours, 26 minutes, and 36 seconds, as against one of 5 hours, 55 minutes, and 10 seconds for the north European station. In other words, there would be a very sensible difference in durations between the two, of 28 minutes and 34 seconds.[10] Incidentally, the island nearest to Pingré's theoretical station for this calculation seems to have been—as we discover from modern maps—Pitcairn Island, where the mutineers of the *Bounty* settled in 1790.

[9] Pingré, "Examen critique des observations du passage de Vénus sur le disque du Soleil, le 3 Juin 1769; et des conséquences qu'on peut légitimement en tirer," Bib. Ste. Gen., Pingré MSS, 2312, fol. 336. A slightly different version of this is available in *Mémoires*, 1770, p. 558.

[10] From the report on Pingré's memoir by Chappe and Lalande, *Proc. Verb.*, 1767, fols. 49-49 verso.

Pingré also discussed the selection of all stations for 1769, including those in Manila and the Marianas, which Le Gentil had suggested. On the basis of his general study, Pingré drew up a special report on Le Gentil's proposals in which, though he agreed with him about the enhanced visibility of the phenomenon in those parts, he nevertheless minimized their value in favor of the observations from the eastern coast of India, preferably from Pondichery again.[11] It was thus at Manila on 10 July 1767, about a year after he had written from the Isle de France, that Le Gentil received the news that his most recent voyage was not looked upon favorably. The letters had apparently taken a fully Spanish route, going to Acapulco, Mexico, and then to the Philippines. "M. de la Lande me fit observer," Le Gentil wrote, "qu'il avoit été lû à l'Académie par M. Pingré, un Mémoire dans lequel il se plaignoit que j'allois trop loin; il auroit voulu que je fusse revenue à Pondichéry. Au reste il étoit assez égale, selon M. de la Lande même, que je restasse à Manille, ou que je revinsse à Pondichéry; & il me dit que c'étoit à mon goût à me décider."[12]

Though it was thereby left up to him to decide, and in spite of his earlier arguments in favor of Manila which he continued to hold, Le Gentil accepted the advice from Paris to return to Pondichery. Perhaps it was merely the suggestion of authority from Paris which Lalande's letter gracefully brought, or perhaps it was due to the assorted troubles Le Gentil was then having with the corrupt Governor of Manila; but for either reason, it was a good time to go. Yet ever mindful of his scientific mission, he left instructions with a certain "Père Théatin, Italien, bon Mathématicien & Missionnaire; il parloit le François," for observing the egress of Venus and regulating his pendulum

[11] Pingré, "Reflexions sur le projet de M. le Gentil à l'égard de l'observation de la prochaine conjonction écliptique de Vénus et du Soleil," Bib. Ste. Gen., Pingré MSS, 2312, fol. 354.

[12] Le Gentil, *Voyage dans les mers de l'Inde . . .* , p. 28. In a later *mémoire*, "Remarques sur le passage de Vénus qui s'observera en 1769," *Mémoires*, 1768, p. 236, but not actually published until 1770, Lalande said otherwise, perhaps influenced by the fate of the Pondichery observation. "Le projet que M. le Gentil a communiqué à l'Académie d'aller observer le passage de 1769, aux îles Mariannes, me paroît préférable à celui de faire cette observation à Pondicheri. . . ."

clock, doing him the additional service of determining the meridian of his station.[13]

By means of a Portuguese vessel bound for Madras, Le Gentil was able to get to Pondichery on 27 March 1768, fourteen months before the transit. There, with the aid of Governor Law, kinsman of the notorious financier, an excellent observatory was constructed on the ruins of the city's fort. With more than enough time to prepare for the transit, Le Gentil carefully began to collect the associate data required for a successful transit observation. These included determining the latitude and longitude of his station, a study of atmospheric refractions at the horizon and elevations of 10, 14, and 45 degrees, an observation of a total lunar eclipse with Governor Law, and the inevitable observations of the satellites of Jupiter.[14] Nights at Pondichery were beautiful and clear throughout the early months of 1769, so much so that there was no scintillation of the stars in his 15-foot telescope, which, though often exposed in the night air in a vertical position for several hours, remained free of humidity. In contrast to his first experience with the English off the Pondichery coast in 1761, he now obtained their fullest cooperation, even to receiving an excellent 3-foot achromatic telescope on loan from Madras.

On the eve of the transit, the weather was perfect, and he observed an immersion of the first satellite of Jupiter, confident of the coming day. "On s'empressoit déjà à me faire des complimens," he wrote, but at seven in the morning, when he was to observe the planet's egress, the sun was hidden behind a cloud. Within less than half an hour the day was again clear and serene. After ten years, he had missed both transits. "C'est là," said Le Gentil resignedly, "le sort qui attend souvent les Astronomes. J'avois fait près de dix mille lieues; il sembloient que je m'avois parcouru un si grand espace de mers, en m'exilant de ma patrie que pour être spectateur d'un nuage fatal, qui vint se présenter devant le Soleil au moment précis de mon observation, pour m'enlever le fruit de mes peines & de mes fatigues."[15] In contrast, he learned that good weather at Manila had resulted in a successful observation there.

[13] *ibid.*, p. 30.   [14] *ibid.*, p. 33.   [15] *ibid.*, p. 35.

It was during his stay at Pondichery following the transit, that Le Gentil began acquiring a knowledge of the Hindu astronomy which he brought back to France. He remained at Pondichery for some time, performing useful experiments with the seconds pendulum, determining refractions of the atmosphere at various seasons and temperatures, and mapping the areas around the colony.[16] While he was there, he received the letters of assistance which he had requested from the Spanish Court two years earlier, and though of reduced value at that late date, they did help him to get back to Europe. Before sailing he also saw Veron, the astronomer who had recently made the famous South Seas voyage with Bougainville, who complained of missing the transit because they had been at sea on that important day.[17]

From Cadiz, where he had debarked, Le Gentil took an overland route to Paris via Madrid and Pamplona, sending his instruments and heavy gear by sea, as Pingré had done before him from Lisbon. Indeed, his return to France over the same mountain route produced a similar reaction: "le 8 [Octobre, 1771], au lever du Soleil, nous passames le crête des Pyrénées; & je mis enfin les pieds en France a neuf heures de matin, après onze ans six mois & treize jours . . . d'absence."[18] But since two years had passed after the transit, he was presumed dead and his heirs were already dividing up his estate, forcing him to undertake expensive legal action to retain his property. With all these tribulations behind him, Le Gentil sought the civilizing influence of domesticity, and married.[19] He had a daughter who made him forget all the misfortunes of the past, Cassini tells us, and the Académie des sciences also helped by obtaining an apartment for his use at the Observatoire de Paris.[20] He died in 1792, but the vacancy so created was not filled. "En 1793 il ne fut plus question de nommer aux Académies, on s'occupa de les supprimer."[21]

[16] *ibid.*, pp. 36-45.   [17] *ibid.*, pp. 49-51.   [18] *ibid.*, p. 77.
[19] J. D. Cassini, *Éloge de M. Le Gentil, membre de l'Académie royale des sciences de Paris*, pp. 27-30.
[20] Oral communication to the author from M. André Danjon, present Director of the *Observatoire de Paris*.
[21] J. D. Cassini, *op.cit.*, p. 34.

I I

Towards the end of 1767, the Académie considered a proposal by Chappe, following Pingré's suggestion, to go to the South Seas to observe the transit.[22] Apparently this had been under discussion since the presentation of Pingré's memoir, for La Condamine had earlier written to a friend at the Court of Spain, Don Georges Juan, sometime Commander of Malta and Ambassador to Morocco, about Spanish support and permission for voyages to both the Pacific islands and California. By October of that same year, Juan informed La Condamine that he expected to meet with difficulty on the Pacific islands project, but that "nous aurons toujours la ressource pour envoyer en Californie tous les observateurs que se présenterons ainsi qu'au royaume de Mexique."[23] That expectation of difficulty was fulfilled, and before the month was over, Juan wrote La Condamine again, to tell him that the Spanish refused to take part in any Pacific expedition. As far as California was concerned, he noted that "les Anglois avoient demandé la même permission pour cette peninsule, et elle leur avoit été réfusée. . . . Cependant le Roi a jugé à propos de prendre un autre parti et de demander lui même des astronomes pour les envoyer. Notre Ambassadeur est chargé de cette demand, et Messieurs de l'académie nommeront les sujets qu'il leur plaira."[24] The French bid thus turned into a Spanish invitation, and Juan added that the king was also planning to have two Spanish naval officers accompany the French academicians and take part in the observations.

By August 1768, it had been decided that Chappe would go to the West Coast of Mexico or to Lower California, in response

[22] Proc. Verb., 1767, fol. 242 verso. This entry in the Procès-Verbaux is relatively modern, having been extracted from the Mémoires secrets de Bachaumont. There is an interesting memoir in Chappe's dossier at the Académie des sciences on this subject. It was not written by Chappe, but argues that he should go to the South Seas for a variety of purposes beyond the transit observation. These were to include new geographic discoveries, the natural history of the islands of the South Seas, and possibly, the test of some of the new marine clocks for determining the longitude at sea. Arch. Acad. Sc., Dossier Chappe, "Sur le voiage de M. l'Abbé Chappe d'Auteroche de l'Académie royale des sciences dans la mer du sud. pour l'observation du passage de Vénus sur le Soleil du 3 Juin 1769," fols. 1-10.
[23] Arch. Acad. Sc., Dossier Georges Juan, "Extrait de deux lettres de M. Don Georges Juan à M. de la Condamine," No. 1.
[24] ibid., No. 2.

to the Spanish invitation. At the same time that this decision was reached, another French expedition to the new world was in preparation. This latter voyage was designed to test the marine clocks of LeRoy and Berthoud, and Pingré, who was to be the chief scientist engaged in this task, was also expected to observe the transit. Originally it was hoped that Pingré's observation would be made from Vera Cruz, Mexico, but in a letter to the Académie, the Duc de Praslin[25] vetoed this idea, pointing out that though it would probably be proper to send another French astronomer with Chappe aboard a Spanish vessel for this purpose, it seemed that the Spanish were touchy about having foreign ships call at their colonial enclaves.[26]

Chappe left Paris on 18 September 1768, accompanied by a servant and three others assigned to assist him: Pauly, *Ingénieur-Géographe*, Noël, an artist, and Dubois, a clock maker. At Cadiz on 17 October, a Spanish fleet was forming for Vera Cruz, but difficulties created by the unexpected size of his party and delays in the fleet's formation led Chappe to obtain special orders for the use of a small ship. And together with MM. Doz and Medina, the promised Spanish naval astronomers, they sailed alone, arriving at Vera Cruz on 8 March 1769 after a seventy-seven day crossing.[27] By 16 May they had safely crossed Mexico in an Indian-led safari, and were located at the tip of the Lower California peninsula near Cape Lucas, in a Spanish mission known today as San José del Cabo.

Chappe's observation of the transit on 3 June 1769 was one of the most complete made. His group was excellently equipped with two achromatic telescopes by Dolland, two quadrants of 3 feet and 1½ feet respectively, and a fine pendulum clock by Berthoud.[28] In addition to the transit itself, Chappe's related observations were also of a high order. He was able to determine the longitude of his position by means of the lunar eclipse of

[25] Cousin to the Duc de Choiseul, he had been given the Ministry of Foreign Affairs in the nominal surrender of that office by Choiseul between 1763 and 1766. See J. F. Ramsey, "Anglo-French Relations, 1763-1770. A Study of Choiseul's Foreign Policy," *University of California Publications in History*, XVII, 3 (1939), 146.

[26] *Proc. Verb.*, 1768, fol. 189 verso.

[27] Chappe, *Voyage en Californie pour l'observation du passage de Vénus sur le disque du Soleil, le 3 Juin 1769. . . .* (Paris: C. A. Jombert, 1772), pp. 11-15.

[28] *ibid.*, p. 70.

18 June 1769, as well as by the more common technique of observing the immersions of the first and second satellites of Jupiter. Shortly after the transit took place, an epidemic disease struck the village and mission of San José, killing about three-fourths of the population, including all of Chappe's assistants except the engineer Pauly. But Chappe continued to observe long after he caught the disease. Indeed, the epidemic was at its height when he made the important observation of the lunar eclipse. However, on 1 August 1769, at the age of forty-one, he too succumbed.[29] Only the engineer Pauly survived, to bring back Chappe's papers, the instruments, and the narrative of their misfortune.

Of the various scientific projects undertaken by Chappe in the new world, not too much can be said. It had been his intention to do a natural history of the region around Mexico City, but this was afterward performed by Don Joseph Antonio de Alzate y Ramyrez, correspondent of the Académie whose work Cassini inserted into his publication of Chappe's *Voyage en Californie.*[30] Chappe had, however, managed to determine the longitude of Vera Cruz and Mexico City and to perform a series of oceanographic temperature experiments on the first half of his journey from Spain to Mexico. Before his departure from France, he had also discussed with Lavoisier the problem of determining the density of water at different places on the globe. And for this purpose Lavoisier provided him with a hydrometer which Chappe labeled an *aréometre*, and with a table of relative densities. Apparently Lavoisier thought well enough of these experiments to attempt to edit and publish them, but no such work was ever produced.[31]

The third French expedition of 1769, already referred to, was not primarily concerned with the transit of Venus. Once again, it was Pingré who boarded ship in the name of a scientific prob-

[29] De Fouchy, "Éloge de M. l'Abbé Chappe," *Histoire*, 1769, p. 171.
[30] Chappe, *Voyage en Californie* . . . , pp. 54-68. Incidentally, Pingré did the French translation of this material from Ramyrez.
[31] *ibid.*, pp. 47-52. See also A. N. Meldrum, "Lavoisier's Early Work in Science, 1763-1771," *Isis*, xix, 2, 56 (June 1933), 350. Lavoisier's purpose was to make hydrometry accurate, and towards this end he devised a lightweight hydrometer. "A copy of it was given to Chappe . . . in order to be used for observing the specific gravity of sea-water."

lem; this time, the determination of longitudes at sea by means of precision timepieces. It seems, though, that Lalande was originally expected to undertake the voyage, but pleaded with the Duc de Praslin to be excused because he could scarcely step aboard an anchored vessel without becoming horribly seasick.[32] It was therefore on 9 December 1768, without Lalande, that the *Isis*, under the command of the Comte de Fleurieu, cast off on her experimental voyage. Besides Pingré and Fleurieu, there were aboard ship two other officers capable of making good astronomical observations, and instruments available for their use. Thus, by 23 May 1769, when they were at Cap-François, Saint-Domingue, preparing to observe the transit, there were four rather skilled observers to do the job.[33] On the day of the transit, they were able to note the external and internal contacts at ingress, and to make several measurements of the least distances between the centers of Venus and the sun. But the position of Cape Francis, which they also determined, did not allow them to witness the entire duration of the transit.[34]

· Apparently undaunted by the low opinion of the Russians which Chappe had expressed, the Imperial Academy at St. Petersburg, with the concurrence of the Empress, again invited the French to send observers to Russia. The invitation was brought up for discussion by the Académie des sciences on 7 December 1768, and a commission consisting of Cassini, Maraldi, and Lalande appointed to consider the qualifications of candidates selected.[35] But this seems to be all the action which the invitation ever stimulated, for no French observers took part in the transit observations of 1769 from Russian soil. The French overseas expeditions in 1769 therefore involved the same personnel that had taken part in similar ventures with regard to the earlier observations. Others had been considered: a certain

---

[32] Pingré, "Voyage aux Isles de l'Amérique," Bib. Ste. Gen., Pingré MSS, 1805, fol. 41. A good part of the material from this Journal went into the published volumes on the voyage. See Comte d'Eveux de Fleurieu, *Voyage fait par ordre du Roi en 1768 et 1769, à différentes parties du monde, pour éprouver en mer les horloges marine inventées par M. Berthoud* (2 vols., Paris: Imprimerie Royale, 1773).

[33] Pingré, "Voyage aux Isle de l'Amérique," *op.cit.*, fol. 75.

[34] Fleurieu, *op.cit.*, I, 124-125; II, 128-239.

[35] *Proc. Verb.*, 1768, fol. 252 verso.

Father Christophe destined for Mexico,[36] Don Georges Juan for California,[37] and, for the Russian voyage, an unknown individual whose qualifications had been weighed and apparently found wanting. But whatever the reasons may have been—unqualified additional candidates, lack of funds, or satisfaction with the experienced observers at hand—French observations abroad, as we have already seen, were in the hands of Le Gentil, Pingré, and Chappe d'Auteroche.

III

Once again, the trace of British activities in the history of the transits of Venus is more distinct than that of any other nation. We are therefore again able to note in detail the character and sequence of events behind their observations of the century's second transit. Thus, as early as 1763, the *Philosophical Transactions* carried James Ferguson's description of the forthcoming transit,[38] and two years later, the much more important memoir by Thomas Hornsby.[39] Hornsby began by reviewing the problem, much in the manner that Pingré did for the Académie in 1767. The transit of 1761 had enclosed the solar parallax between limits of 8.5 and 10 seconds, but this was an unsatisfactory variation for so small a figure. "In this uncertainty," he wrote, "the astronomers of the present age are peculiarly fortunate in being able so soon to have recourse to another transit of Venus in 1769, when, on account of that planet's north latitude, a difference in the total duration may conveniently be observed, greater than could possibly be obtained, or was even expected by Dr. Halley, from the last transit."[40]

[36] *ibid.*, 1767, fol. 289 and 1768, fol. 1. Christophe was a Capuchin monk who had submitted a paper to the Académie on the coming transit and its observation from Mexico. Cassini de Thury, Jeaurat, and Le Monnier were appointed to evaluate it, and did so favorably, so that for a while it seemed as if he might be appointed to make the observation there.

[37] Don Georges Juan was the Spanish correspondent to whom La Condamine had written about the possibility of getting Spanish aid for an expedition to the South Seas or California.

[38] J. Ferguson, "A Delineation of the Transit of Venus expected in the Year 1769," *Phil. Trans.*, 1763, p. 30.

[39] T. Hornsby, "On the Transit of Venus in 1769," *Phil. Trans. Abgd.*, xii, 265-274.

[40] *ibid.*, p. 265.

We have seen Pingré echo this same thought, arguing for the method of durations in '69 as the most promising of all available techniques. After indicating that this emphasis was a valuable reward of the experience of 1761, Hornsby then turned to a discussion of stations that would, in 1769, yield observations of maximum and minimum durations. And this singular interest immediately brought him to focus on the islands of the South Seas for possible stations. Accounts of the voyages of the sixteenth and seventeenth centuries were examined for all references to such places, their probable locations, and the conditions of environment and population connected with them. Hornsby thus considered the Spanish voyages of Alvarez de Mendoza and Juan Fernandez de Quiros to the Solomon Islands, and the Dutch explorations of Abel Tasman in Australian waters. From these and the best of contemporary French and British maps, he drew up a list of seventeen island groups between south latitudes of 4 and 21 degrees, and west longitudes (from Greenwich) of 130 and 190 degrees.[41]

Into this rather large expanse of Pacific waters, where there was a relative certainty of land, Hornsby invited the interested European powers to send their expeditions. "Posterity must reflect with infinite regret," he warned, "their negligence or remissness; because the loss cannot be repaired by the united efforts of industry, genius, or power." But lest this appeal to the opportunity of a moment, in the name of science and national honor, or to the judgment of posterity fall upon cynical minds and unwelcome ears, Hornsby further added that it is a "worthy object of attention to a commercial nation to make a settlement in the great Pacific Ocean."[42] In the 1769 quest for Venus in transit then, the issues of geographical exploration and commercial exploitation were joined to those associated with the transit of 1761—the improvement of navigation and the study of natural history.

[41] *ibid.*, p. 271.

[42] *ibid.*, p. 274. In a brilliant introduction to the newest edition of Captain Cook's Journals, Beaglehole sums up both the commercial and the scientific traditions, at mid-century, which encouraged the dispatch of voyages of exploration. J. C. Beaglehole (ed.), *The Journals of Captain James Cook On His Voyages of Discovery, The Voyage of the Endeavour 1768-1771* (Cambridge: Hakluyt Society, 1955). See especially pp. lxxiv-lxxxiii.

It was thus on 5 June 1766, four years after Hornsby's paper but early enough for adequate preparation, that the Council of the Royal Society met to discuss the practical arrangements for observing the coming transit. In the final moments of that discussion, it resolved unanimously to supplement whatever the British themselves might directly do, by engaging an additional astronomer from abroad to observe under the Society's auspices. With characteristic resolution, the Council immediately selected Father Roger Boscovich, then Professor of Mathematics at the University of Pavia and a foreign Fellow of the Royal Society. At the same time, an order was issued recalling all the instruments which had been manufactured for the first transit from the individuals with whom they had been deposited for protection and use during the intervening years.[43]

More than a year went by before the Council again discussed the transit of Venus. In November 1767, a special Committee on the Transit of Venus was formed, consisting of Campbell, Cavendish, Short, Bevis, Raper, Maskelyne, Murdoch, Ferguson, or any three of them, to decide upon the persons, places, and methods to be employed in the Royal Society's official observations of the transit.[44] In practice, the scientific work was actually performed by Maskelyne, Bevis, Short, and Ferguson, though the entire Committee took part in the vote on final decisions. A special meeting of this group took place on 17 November, at which time the four astronomers just mentioned read individual papers in keeping with the purpose for which the Committee had been created. The final report embodying the essential features of each of these was read to the Council by Maskelyne two days later.[45]

The general conclusions which the Committee had reached now excluded Boscovich and the California observations, but did include the following agreements:

1. It would be proper this time to send observers to Fort Churchill on the western side of Hudson's Bay; and arrangements should be made with the Hudson's Bay Company for transportation to and from the station and for assistance at the site.

[43] CMRS, V, 155-156.   [44] ibid., p. 184.   [45] ibid., p. 187.

2. Observations should be made at Vardö, a small island off the northern Finnish coast, and at North Cape, the northernmost tip of the Scandinavian peninsula, by British observers unless it was learned that Swedish or Danish astronomers were planning to make use of these stations. In any case, the Admiralty was to be consulted on the use of the annual ship sent out to those waters to protect British northern fisheries.

3. At least two observers should be sent to the South Seas to find an island station for the transit observations somewhere in the general area already suggested by Thomas Hornsby. For this purpose the government was to be asked "to furnish a Ship to attend . . . upon this Business."

4. Each observational team should be equipped with a quadrant, a clock, two reflecting 2-foot telescopes with micrometer attachments, and a thermometer, barometer, and compass.

5. The men best suited and available for the task were Bradley (nephew of the former Astronomer Royal), Mason, Dixon, Dunn, Green, Dymond, Wales, and Dalrymple, with the latter kept particularly in mind for the South Seas voyage.

Several other decisions were reached in this report. The Committee resolved to write to Sweden to discuss the northern observations; it drew up a set of instructions for observing the transit to be distributed by the East India Company among its plantations; and it also determined the optimum dates of departure for the Hudson's Bay team and those destined for the South Seas.[46]

Profiting by the lessons of 1761, Bevis made the interesting suggestion in his report that the observers selected be trained in transit observations by means of a mechanical model. This was to consist of an artificial sun across which a miniature Venus would transit, the observers making identical observations with those they would engage in on 3 June 1769, especially measurements of the least distances between the centers of the planet and the sun.[47] But there is no way of knowing whether or not this scheme was ever put into practice. The meeting at which

[46] ibid., pp. 193-207. These pages contain the four reports from which these common conclusions were drawn, having been recorded in the Council Minutes for 30 November 1767 by a vote of the Council.
[47] ibid., p. 201.

this suggestion was made also reveals something of the manner in which men were recruited for the expeditions. For Maskelyne wrote to the universities—to Hornsby at Oxford, and Shepherd and Ludlam at Cambridge—asking them to recommend likely candidates for the voyages to foreign parts.[48]

By December 1767, the Council had set in motion the necessary procedures for implementing the above resolutions. Throughout the month, interviews were held with the men named as possible observers in the Committee report, and dispatches were sent off to appropriate participating agencies, the Crown, the East India Company, the Hudson's Bay Company, and so forth. Salaries and other expenses were discussed, as well as the special conditions, if any, proposed by those nominated, such as Dalrymple's requirement, to wit: "I can have no Thoughts of undertaking the Voyage as a Passanger going out to make the Observation, or on any other footing, than that of having the management of the Ship intended for the Service."[49]

Thereafter, the final decisions were made. By January 1768, Dymond and Wales were informed that they were selected for the Hudson's Bay station; that they were to leave the following May; and that upon their return, they would receive a payment of £200 each.[50] Towards the end of the month, the Society was also informed of the Hudson's Bay Company's cooperation, which included recommendations that the building intended for the observatory be prefabricated in England because of the shortage of proper wood in the region.[51] A special octagonal building was then designed by Smeaton, Maskelyne, and Ashworth, the Greenwich carpenter. It included brass rings and rollers for a slotted roof, so that it was a true portable observatory, very likely the first of its kind.[52]

Besides those available from the previous transit, additional instruments had to be acquired to fulfill the broadened opera-

[48] *ibid.*, p. 207.   [49] *ibid.*, p. 236.   [50] *ibid.*, p. 268.
[51] *ibid.*, pp. 281-284.
[52] *ibid.*, pp. 290 and 300-301. The instruments to be used by Dymond and Wales at Hudson's Bay were also enumerated here, to wit: Ellicott's clock; telescope by Short with Dolland's micrometer; another clock by Shelton; an alarm clock, barometer and two thermometers designed to read well below the freezing point Fahrenheit, by Bird; as well as a 1-foot astronomical quadrant by Bird, a variation compass and an additional Hadley's quadrant.

tions of the Royal Society in 1769. This meant an order for 6
quadrants, 4 clocks, 2 reflecting telescopes, 5 telescope stands,
3 barometers, 3 thermometers, 1 Dolland micrometer, and 1
dipping needle—all of it to come to £293.[53] With some idea of
expenses now in mind, the Council drew up and signed on 15
February 1768 the Society's petition to the King for help. It
was similar to the memorial presented on behalf of the 1761
transit, making its appeal in terms of aid to navigation, national
honor in the face of parallel activities by other nations, the
rarity of the phenomenon and the priority of English interest in
it, and the greater advantage of this second transit over the first.
In support of that contention, they emphasized the importance
of an observation from the southern latitudes of the Pacific
Ocean. Finally, the expected important conclusion was drawn
that the expenses for this large undertaking, exclusive of the
shipping involved, came to about £4,000, an amount well beyond
the Society's means.[54] Yet without much delay, the Crown
honored the Society's request to the full amount on 24 March
1768.[55]

Then, doubly informed by the King as well as the Royal
Society, the Admiralty initiated its own activities on behalf of
the South Seas program. A cat-built, 370-ton former coal bark
was purchased and refitted for the voyage, and the Society was
so informed in April.[56] But to the Council's proposal that
Alexander Dalrymple be given command of the vessel, the Lords

[53] ibid., p. 286.   [54] ibid., pp. 292-295.
[55] ibid., p. 303. This proved to be more than was needed for the entire operation, and the excess of funds led to an interesting by-product of the transit story.
When Dymond and Wales returned from Hudson's Bay, they had spent less
than had been anticipated. The Royal Society then asked the King what he
desired done with the surplus funds, and he informed them that they should
employ it in any experiment that they deemed important. After giving much
thought to the matter, they used the money to finance the famous Schehallion
experiment by Maskelyne, to determine the density of the earth. See Charles
Hutton, "Lettre du Dr. Hutton à M. le Marquis de Laplace," *Journal de physique, de chimie, d'histoire naturelle et des arts*, xc (Avril 1820), 308.
[56] A "cat-built" ship meant one that was either built in Norway (cat stems
from the Norwegian word for ship), or in Yorkshire ports where the construction type was imitated. These were roomy, comfortable ships of shallow draft,
better suited for the voyage than the man-of-war which was originally considered.
The purchase price was £2800, and it cost £2294 to refit her. G. A. Wood,
The Voyage of the Endeavour (Melbourne: Macmillan and Co., 1944), p. 50.

of the Admiralty turned a deaf ear. And on his part, Dalrymple persisted in maintaining his earlier attitude, so that the Society was forced to drop him.[57] Obviously the objection to Dalrymple was not based on his fitness for the task, for as an agent of the East India Company, he had sailed in the Eastern oceans and knew them well. Moreover, Dalrymple was an able geographer and navigator, who during this very period had been trying to persuade British statesmen about the existence of a continent in the southern latitudes worth discovering and annexing. But the reef upon which Dalrymple's commission foundered was the naval regulation requiring the commander of a King's ship to be an officer of the King's Navy.[58] The result was that the Admiralty named a young, unknown Lieutenant to the post. On 5 May 1768, James Cook was presented to the Council as the Captain of the bark *Endeavour*.[59] At the same meeting, Charles Green, former assistant at the Greenwich Observatory, was also appointed a member of the expedition.

Interest in the voyage to the South Seas was not limited to the transit of Venus alone, for one of the greatest ambitions of the young King (George III was twenty-eight in 1768) was to achieve the exploration of the unknown areas of the globe. Relatively few explorers had penetrated south of the fiftieth parallel in either the Atlantic or the Indian Oceans; frozen seas had brought a halt to the hoped-for discovery of the Northwest Passage, and though many had entered into its pellucid waters, the Pacific Ocean remained almost as unknown as it was vast.[60] But to the motive of geographical discovery and empire, one can add still another, symbolized by the presence of Joseph Banks, "a Gentleman of large fortune who is well versed in

---

[57] CMRS, V, 305-307.

[58] Wood, *op.cit.*, pp. 32-34. Whatever else lay behind this regulation, the Lords of the Admiralty remembered all too well the failure of the *Paramour Pink*, under the command of Edmund Halley, to accomplish her mission. Sent out to study the variation of the compass at sea in 1698, the ship was forced to return because Halley could not properly control the crew. Beaglehole, *op.cit.*, p. cv.

[59] CMRS, V, 314. Captain Cook received "the Sum of One Hundred Guineas as a Gratuity for his Trouble, as one of the Observers in the Southern latitude," *ibid.*, p. 322.

[60] H. Carrington (ed.), *The Discovery of Tahiti* (Publications of the Hakluyt Society, Series II, XCVIII; London: The Hakluyt Society, 1948), xix.

natural history,"[61] and his suite of six assistants aboard the *Endeavour* when she sailed to southern waters. This part of the expedition, to explore the natural history of the South Pacific, went to sea far better equipped than any other similar venture up to that time. Besides artists and the botanist—Dr. Solander, pupil of the great Linnaeus—there were a wealth of equipment and a complete library of natural history, so that it was estimated by a contemporary that the cost to Banks must have been about £10,000.[62]

The fame of both Captain Cook and Joseph Banks, who was eventually to leave an indelible imprint upon the Royal Society as one of the most forceful presidents in its history, is so widespread as to preclude comment in this study. They sailed from Plymouth on 26 August 1768, on a voyage which lasted until 1771 and reduced their original complement of men from ninety-four to fifty-six, but which was successful in the extreme and brought them world-wide fame.[63] Except for the theoretical limits established by the Astronomer Royal, Cook's original instructions required only that he be somewhere in the South Seas by June 1769. But only a few weeks before he sailed, Captain Samuel Wallis and H.M.S. *Dolphin* returned from the Pacific to announce the discovery of Tahiti. This was most opportune, for it was now possible to appoint a new goal, a place where the expedition might receive a friendly welcome and be certain of adequate food and water and, above all, of a secure place of observation. And on 9 June 1768, the Society indicated its desire to the Admiralty that Tahiti be fixed as the expedition's destination.[64]

In addition to these places the Society continued to weigh the dispatch of other expeditions. Throughout the summer of 1768, from July to September, consideration was given to both

[61] CMRS, V, 335.

[62] Wood, *op.cit.*, p. 49. A work on this aspect of the Cook voyage, with emphasis on anthropology, is in preparation by D. S. Marshall of the Peabody Museum, Salem, Massachusetts; letter to the author dated 1 May 1954. Additional material of the same nature is to be contained in the final volume of Cook's Voyages edited by J. C. Beaglehole.

[63] W. J. L. Wharton (ed.), *Captain Cook's Journal During His First Voyage Round the World* (London: Elliot Stock, 1893), lii-liv.

[64] CMRS, V, 335.

Spitzbergen and North Cape, Maskelyne suggesting that both might be used and that his present assistant, William Bayley, would be willing to go to either place.[65] Mason and Dixon were again invited to take part in the transit observations, but only Dixon expressed a willingness to go north. The result was that by the end of December it had been decided that Dixon and Bayley would travel north together but observe at different places, Bayley at North Cape[66] and Dixon at Hammerfest,[67] an island off the Norwegian coast about eight miles southwest of North Cape. In the meantime, at Maskelyne's request, Mason agreed to go to County Donegal in northwest Ireland to observe the transit.[68]

For all of these expeditions (properly speaking, four in number), the kind of preparations were in order that we have by now come to expect. Shipping had to be arranged for the North Cape expedition (the Admiralty provided a Captain Douglas and H.M.S. *Emerald*);[69] and, where necessary, permission had to be obtained from the proper authorities (the Danish King informed the Society of his full cooperation on 6 April 1769).[70] And for all expeditions wherever bound, separate sets of detailed instructions, including all the observations expected of them, were drawn up by Nevil Maskelyne in his official capacity as Astronomer Royal.[71] However, in the last few months before the transit, the Royal Society received two interesting communications upon which it did not act. The first of these was an offer from the Elector Palatine to send, at his own expense, his official astronomer, Father Christian Mayer, F.R.S., wherever the

[65] *ibid.*, pp. 353-356.
[66] W. Bayley, "Astronomical Observations made at the North Cape, for the Royal Society," *Phil. Trans.*, 1769, 262-272. As a by-product of their stay in the North, Bayley and Dixon jointly produced "A Chart of the Sea Coast and Islands near the North Cape of Europe."
[67] J. Dixon, "Observations made on the Island of Hammerfost, for the Royal Society," *Phil. Trans.*, 1769, pp. 253-262.
[68] CMRS, V, 367, and C. Mason, "Astronomical Observations made at Cavan, near Strabane, in the County of Donegal, Ireland, by Appointment of the Royal Society," *Phil. Trans.*, 1770, pp. 454-496.
[69] CMRS, VI, 20.
[70] *ibid.*, p. 19.
[71] For Dymond and Wales on 9 June 1768, CMRS, V, 329-334; for Cook and Green on 13 June 1768, *ibid.*, pp. 339-346; for Mason, Bayley, and Dixon on 21 April 1769, CMRS, VI, 30-42.

Society would like to have him go to observe the transit.[72] The second was a request from the Grand Duke of Tuscany to allow him to send a young Tuscan astronomer to accompany the Royal Society's northern expedition. This was refused on the rather flimsy ground that arrangements with the Admiralty were already completed.[73] But the reasons for turning down the generous offer in the first case do not appear anywhere in the documents. Of the observations themselves, all were successful, and the results will be given in tabular form in the pages to follow.

I V

Another phase of British overseas activities with regard to the transit of 1769 is also worth detailing. This is the independent decision to observe the phenomenon undertaken in the North American colonies. For one recent writer, American participation in the transit observations of 1769 was a "symbolic act of allegiance to science and learning,"[74] but this is a patriotic turn of phrase a little too strongly put, for the viability of American science in the eighteenth century was not as fully based on the practice of astronomy as that language implies. Nevertheless the subject was popular, and within its framework, the transit of Venus was rare and exciting enough to produce a wide-scale set of observations, multiplying Winthrop's unique observation of 1761 many times. If the observations made in the new world by expeditions sent from the old are omitted, and if, of the remaining number, only those are counted which entered into the mainstream of scientific literature available to the world at large, then there were nineteen observations made in the British colonies of North America[75]—a very respectable number.

Newspapers and almanacs publicized the event, though not always with the greatest accuracy, as, for example, Abraham

[72] CMRS, V, 367.

[73] CMRS, VI, 6.

[74] D. Fleming, *Science and Technology in Providence, 1760-1914* (Providence: Brown University, 1952), p. 13.

[75] J. F. Encke, *Der Venusdurchgang von 1769 als Fortsetzung der Abhandlung über die Entfernung der Sonne von der Erde* (Gotha: Becker'schen Buchhandlung, 1824), p. 64. An additional observation cited by Encke that might be of independent American origin, is Alzate's made in Mexico, but this was obviously not British and I have omitted it from the present count.

Weatherwise's *New-England Town and Country Almanac* for 1769, which stated that the transit would be invisible in Britain and America.[76] Amateurs of all sorts took an active interest in the observations, and some of these recorded their data in diaries or private journals, so that they were not included in the nineteen observations referred to above. Important men in colonial affairs like Manassah Cutler[77] and Ezra Stiles[78] belong in this group. But for the major observations we must turn to the professionals and the organizational pattern of their actions.

Once again John Winthrop took a leading hand in transit affairs. Upon the basis of his successful experience in procuring an observation of the transit of 1761 from Newfoundland, Winthrop wrote to James Bowdoin on 18 January 1769 for similar assistance. But he did so upon the stimulus of a letter from Franklin, whose words Winthrop quoted to Bowdoin as follows:

M. Maskelyne . . . wishes much that some of the Governments in North America would send an astronomer to Lake Superior to observe this Transit. I know no one of them likely to have a Spirit for such an undertaking, unless it be the Massachusetts. . . . If your health & strength were sufficient for such an expedition I should be glad to hear you had undertaken it. Possibly you may have an Eleve that is capable. The fitting you out to observe the former Transit, was a Public Act for the Benefit of Science, that did your Province great honor.[79]

To this request, now addressed to Bowdoin as well, Winthrop added various positive arguments that might appeal to a Provincial Government, such as the advantages to be gained from an exploration of the unknown regions around the Great Lakes, and the new and useful maps which would thereby be produced for trade routes. Although Winthrop felt himself to be unfitted for the task because of poor health, he knew of many accomplished young men capable of undertaking the trip. Finally he

[76] A. E. Lownes, "The 1769 Transit of Venus and its Relation to Early American Astronomy," *Sky and Telescope*, II, 6 (April 1943), 4.

[77] W. P. Cutler (ed.), *Life, Journals and Correspondence of Rev. Manassah Cutler, LL.D.* (Cincinnati: R. Clarke and Co., 1888), I, 20.

[78] F. B. Dexter (ed.), *The Literary Diary of Ezra Stiles* (New York: Scribner's & Sons, 1901), I, 12-13, 15.

[79] J. Winthrop, Massachusetts Historical Society (hereafter cited: *Mass. Hist. Soc.*), Winthrop Papers, XXII, 5, Letter to James Bowdoin, Cambridge, 18 January 1769, fol. 1.

suggested that if Bowdoin and Governor Temple would use their influence with General Gage, those selected for the trip might receive the General's permission to travel with one of the military convoys delivering supplies to the Lake forts.[80]

This letter was the first in a long triangular correspondence between Bowdoin, Gage, and Winthrop. By 23 January 1769, Bowdoin had written Gage to repeat the essentials of the above letter, and to enlist his help in the transit cause. The same day, he optimistically informed Winthrop that "I have hopes the Genl will make the expense of it a contingency within his own department."[81] But when Gage replied about a week later, there was every indication of cooperation save that of financial support. Even more importantly, Gage informed Bowdoin that "some Gentlemen from Philadelphia made Application to me some Months ago, concerning the like Intentions of sending some Astronomers . . . to Lake Superior. . . . Perhaps they would be glad to join those from Boston."[82] The exchange of letters covering a possible cooperative effort between Pennsylvania and Massachusetts ran through March of 1769, but the entire enterprise collapsed when the Governor and Council of Massachusetts voted not to support the expedition, and Gage could not find the means to do so.[83] Philadelphia partisans of the Lake Superior station had apparently fared no better in their independent attempt to get an expedition into the American Northwest, for no observation of the transit was ever reported from that region.

Winthrop thus prepared to observe the transit in Cambridge as best he could. But he also tried to multiply his own efforts as much as possible and, at the same time, to inform the general public of the reasons behind the importance of a transit of Venus. To this end, he gave two public lectures in Holden Chapel at Harvard University on the connection between the transits of

[80] *ibid.*, fol. 2.
[81] Collections of the Massachusetts Historical Society (hereafter cited: *Col. Mass. Hist. Soc.*), 6th Series, IX, The Bowdoin-Temple Papers, 120.
[82] *ibid.*, p. 121.
[83] *ibid.*, p. 130. General Gage actually suggested that a subscription be taken up to support the enterprise, and though it seemed likely that the money could be so raised, the late date prevented the execution of the plan.

Venus and the solar distance. The tone and quality of Winthrop's popularization can best be determined from this brief excerpt from the first lecture.

On account of their rarity alone [the transits of Venus], they must afford an exquisite entertainment to an astronomical taste. But this is not all. There is another circumstance which strongly recommends them. They furnish the only adequate mean of solving a most difficult Problem,—that of determining the true distance of the Sun from the Earth. This had always been a principal object of astronomical inquiry. Without this, we can never ascertain the true dimensions of the solar system and the several orbs of which it is composed, nor assign the magnitudes and densities of the Sun, the planets and comets; nor, of consequence attain a just idea of the grandeur of the works of GOD.[84]

Upon this kind of foundation, Winthrop elaborated the principles behind the transit observations, and, in a special appendix to the printed version of the lectures, the best techniques for actually performing them.

In Providence, Rhode Island, a wealthy merchant, Joseph Brown, "a gentleman of solid, active genius, strongly turned to the study of mechanics and natural philosophy,"[85] became interested in observing the transit of '69 as a result of reading Winthrop's pamphlet on the transit of '61. Following Winthrop's account of the instruments required, he sent to London for a telescope, and afterwards consulted with Benjamin West, local almanac author, bookseller, and mathematician. From West, he learned of instrumental improvements that had been suggested by the American Philosophical Society in their own application for funds for transit operations from the Pennsylvania Assembly. The usual observations associated with a transit were begun about a month before the transit itself, under Benjamin West's direction, who tells us of one interesting aspect of the determination of the meridian. "That our observations might be as useful as possible," he wrote, "notice was given beforehand to the people . . . that on the day before the transit, when the sun

[84] J. Winthrop, *Two Lectures on the Parallax and Distance of the Sun, as Deducible from the Transit of Venus* (Boston: Edes & Gill, 1769), p. 14.
[85] B. West, *An Account of the Observation of Venus Upon the Sun, the third day of June, 1769 at Providence, in New England* (Providence: John Carter, 1769), p. 10.

came on the meridian, a cannon would be fired, which being done, most of the inhabitants marked meridian lines in their windows, or on their floors."[86] The transit itself was successfully observed on 3 June 1769 at an observatory erected on a hill near the present Transit Street in Providence (Figure 12).[87]

The major set of American-sponsored observations were those that received their support from the American Philosophical Society at Philadelphia. As early as April 1768, a movement was set afoot within the organization to sponsor such observations, and the permanent Committee for Natural Philosophy and Astronomy was called upon to draw up an estimate of the probable expenses for such an undertaking in the Philadelphia region.[88] Two months later, projections of the transit of 1769 by David Rittenhouse and John Ewing were presented at a meeting of the Society. At the same session committees were also appointed to prepare for observations at Norriton and Philadelphia, the Society offering to defray the expenses of both operations.[89]

But the passage of time also heightened their ambitions; and on 20 September 1768, James Dickenson proposed to the Society that they support an expedition to the North Country, "for observing the ensuing Transit of Venus . . . and for reconnoitering and making a map of the Country from the South End of Hudson's Bay and to extend thence towards the Head of the Mississippi [*sic*]."[90] Obviously this is the parallel suggestion to Winthrop's desire for a Lake Superior observation and concomitant explorations, though it is much more specific and implies a large-scale operation. Appeals for this purpose were addressed to General Gage in New York by both groups, as we have already noted, but the Pennsylvania Assembly, which was called upon by the Society to support this enterprise, did not see fit to do so,

[86] *ibid.*, pp. 14-15.
[87] Lownes, *op.cit.*, p. 5.
[88] *Early Proceedings of the American Philosophical Society . . . Compiled by One of the Secretaries from the Manuscript Minutes of Its Meetings from 1744 to 1838* (Philadelphia: McCulla & Stavely, 1884), p. 14. But no such estimate can be found in these proceedings or the manuscript originals.
[89] *ibid.*, p. 15. Smith, Rittenhouse, Lukens, and Dickenson were selected for the Norriton station (actually Rittenhouse's farm), and Ewing, Williamson, Joseph Shippen, and Prior for the Philadelphia observations. Incidentally, Rittenhouse is another example of an instrument maker who observed the transit.
[90] *ibid.*, p. 18.

save to grant £100 for the purchase of a reflecting telescope in England.[91] But this narrow vision of both the Massachusetts and Pennsylvania Assemblies in failing to stand behind a broad scientific expedition into the Northwest territories would some day be rectified. Years later, as united states rather than separate provinces, they would find themselves indirectly lending support to such a venture in Jefferson's imaginative action dispatching of the Lewis and Clark expedition.

Nevertheless, between the Pennsylvania Assembly, the American Philosophical Society, the Library Company of Philadelphia, and the interest of several private citizens, it was found possible to support three stations of observation (the third was at Cape Henlopen, Delaware)[92] and supply them with adequate instruments. Although Franklin, the Society's president, was abroad at the time of the transit (indeed, he was first to preside over a meeting of the Society on 27 September 1776, after almost seven years as its president),[93] he did as much as possible to aid the Philadelphia group, putting them in contact with Maskelyne and other British scientists and helping them to procure instruments.[94] The transit was very well observed, and the results when sent abroad were found to be highly satisfactory. Maskelyne, for example, thought them "excellent and compleat . . . [they] do Honor to the Gentlemen who made them. . . ."[95] These observations, together with those made independently from other stations in America will be recorded in the tabular summary to follow.

[91] *ibid.*, p. 19.

[92] *ibid.*, p. 37. This was manned by Owen Biddle, to whom the Library Company of Philadelphia lent its large reflecting telescope, *ibid.*, p. 39. Biddle was assisted by Joel Bayley, "who was a considerable time with Dixon and Mason running the line between Pennsylvania and Maryland," Archives of the American Philosophical Society, Franklin Papers, II, No. 180, Letter from Cadwalader Evans to Benj. Franklin, dated 11 June 1769.

[93] C. Van Doren, *Benjamin Franklin* (New York: Garden City Publishing Co., 1941), pp. 535-536.

[94] *Arch. Am. Phil. Soc.*, Franklin Papers, II, No. 179, Letter from Thomas Bond to Benj. Franklin dated 7 June 1769. ". . . [W]e are much obliged for your Care in procuring the Telliscope, which was used in the late Observations of Venus's Transit. . . ."

[95] Letter from Maskelyne to Thomas Penn read into the minutes of the 15th December 1769 meeting of the American Philosophical Society, *Early Proceed. Am. Phil. Soc.*, *op.cit.*, p. 46.

V

In keeping with the 1769 emphasis on the method of durations as the most likely to prove successful in the search for a definitive solar parallax, the stations of northern Europe assumed a position of major importance. For it was expected that those observations from this region, especially from Lapland, would be of the necessary lengthened duration to match the shortened duration of the South Seas station. The British dispatch of Dixon and Bayley to Hammerfest and North Cape was designed to partially meet the requirement of data from this area. But of even greater importance was the station on the island of Vardö, off the northeastern coast of Norway, at 30 degrees, 7 minutes east longitude and 70 degrees, 22 minutes north latitude. Considering the position of the island, it has known many important visits. It contained a fortress, Wardhus, built about 1310, from which the transit observations were made; it had been visited by Scandinavian rulers through the centuries and, in our day, by the explorer Fridtjof Hansen in the last decade of the nineteenth century.[96] And it was under the sponsorship of Christian VII, King of Denmark and Norway, that an invitation was extended to Father Maximilian Hell to lead an expedition to that island.

With the permission of the imperial government of Austria-Hungary and the superiors of his order, Hell left Vienna on 28 April 1768, accompanied by a fellow Jesuit, Father Johann Sajnovics.[97] On 1 June 1768, they were presented for the first time to their benefactor, the twenty-two-year-old Danish King at his castle near Lübeck. There they discussed the various projects which Hell planned to undertake at Vardö above and beyond the transit observations, and "der König schien mit dieser Äusserung sehr zufrieden."[98] When they left Copenhagen, crossing into Sweden at Hälsingborg for the overland trip to Norway,

[96] Sarton, op.cit., p. 100.

[97] Johann Sajnovics, "Tagebuch der Reise von Wien nach Wardoe und zurück," in C. L. Littrow, P. Hell's Reise nach Wardoe bei Lappland und seine Beobachtung des Venus-Durchganges im Jahre 1769 (Vienna: Carl Gerold, 1835), p. 87. This is a German translation which Littrow made from the original Latin diary in the Archives of the Vienna Observatory.

[98] ibid., p. 100.

they were accompanied for a while by the Danish astronomer Christian Horrebow, who had observed the first transit.[99] By the end of July they were in Trondheim, and shortly after their arrival, a young botanist named Borgrewing, "der in Schweden Linné's Vortrag durch ein ganzes Jahr gehört hatte," was attached to their expedition to collect plants and algae in Lapland and to acquire some astronomical skills as an assistant in the transit observations.[100] On 22 August they sailed from Trondheim, to arrive at Vardö on 11 October;[101] and before the month was over, they had begun observing. Though the skies were often cloudy, on the day of the transit they were lucky enough to observe the contacts at ingress and egress, and this they loudly celebrated afterwards with the firing of cannon and a *Te Deum laudamus*.[102]

The Hell expedition left Vardö for Copenhagen on 27 June 1769, where Hell and Sajnovics stayed for eight months, finally returning to Vienna more than two years after they had departed. The long stay at Vardö had given Hell time to collect a wealth of data on arctic geography, meteorology (including a study of the aurora borealis and barometric fluctuations), anthropology, and the language of the region. He planned to publish an encyclopedic study on the arctic, and went so far as to give it a title, *Expeditio litteraria ad Polum Articum*, and to issue advance publicity on the work in publications like the *Journal des Sçavans*.[103] But in spite of a subsidy from the Empress Maria Theresa, the patronage of the Danish King who had sponsored the project, and the accumulation of many notes, the book was never produced. Indeed, the project was so grandiose in conception, that it would have taken much more than eight months of residence in the arctic to become an expert on the subject, and probably would have required the cooperation of many more scholars.

When the transit was over, Hell delayed for a long time reporting his data to foreign astronomers. Lalande, who was Hell's

---

[99] *ibid.*, p. 103. See Chapter IV as well.   [100] *ibid.*, p. 116.
[101] *ibid.*, p. 125.   [102] *ibid.*, p. 139.
[103] Sarton, *op.cit.*, p. 100. The prospectus in the *Journal des Sçavans* is contained in the volume for 1771, pp. 499-500, reprinted by Sarton, *op.cit.*, p. 104.

official correspondent at the Académie, requested the results of the Vardö observations and was refused. This led him to suspect that Hell had none to give.[104] The story was given further currency by anti-Jesuits, who saw in the incident an opportunity to discredit the Order, then close to dissolution. But on 24 November 1769, the observations were presented to the Danish Academy of Sciences, and even before Hell had returned to Vienna, 120 copies of that report were printed in Denmark, Hell himself retaining 20.[105] The publication of the data served to quell the charges and the story was soon forgotten.

But in 1835, Carl Ludwig Littrow, one of Hell's successors at the Vienna Observatory, discovered the astronomical diary which Hell had kept at Vardö, and published a critical examination of it together with the narrative of the voyage which had been kept by Sajnovics. This reawakened earlier insinuations about Hell's work, and Littrow further charged that Hell had delayed publication of his data in order to produce results which would be in the best possible agreement with the observations of other astronomers. As a matter of fact, he came to the conclusion that erasures had been made of the original entries in the data book and new ones later inserted, often in ink of another color.[106] Years later, in 1883 to be exact, in the course of a visit to the Vienna Observatory, partly to see its new telescope and partly because he was then preparing his own study of the eighteenth-century transits, the American astronomer, Simon Newcomb, examined the Hell documents in the light of Littrow's charges. His conclusions were startling. Erasures and corrections had indeed taken place, but Newcomb was able to prove with complete satisfaction, that they "were made at the time of writing, and without the slightest intention of giving but the actually observed moment when *Venus* was first seen."[107] New-

[104] *ibid.*, p. 100.

[105] Sajnovics, *op.cit.*, p. 158. The pamphlet bore the title *Observatio transitus Veneris ante discum solis die 3 Junii anno 1769, Wardoëhussii, auspiciis potentissimi ac clementissimi regis Daniae et Norvegiae, Christiani VII* (Hafniae: Gerhard Giese Salicath, 1770).

[106] S. Newcomb, "On Hell's alleged falsification of his observation of the transit of Venus in 1769," *Monthly Notices of the Royal Astronomical Society*, XLIII, 7 (May 1883), 371-372.

[107] *ibid.*, p. 376.

comb also demonstrated that what Littrow had seen as differently colored ink was due to a flaw in his own vision: there was evidence at the Vienna Observatory from Littrow's work to show that he was color-blind![108] So it was the nineteenth century which finally saw the vindication of Father Hell, to which George Sarton has added a supporting echo in the twentieth.

Russian interest in the second transit of Venus was even greater than it had been in the first. Informed of the coming transit, Catherine the Great ordered the Imperial Academy, in the spring of 1767, to begin preparations for its observation "with the greatest care." Her brief letter to Count Vladimir Orlov, Director of the Academy, is of the greatest interest in this respect for it reverses the usual action in which appeals were made by scientists to the sovereign, and much of it is worth quoting. Catherine's desires were neatly enumerated:

1) That the Academy . . . make these Observations with the greatest care; and that I desire, in consequence of this, to know 2) what are the places of the Empire the most advantageously situated, and what are those which the Academy appoints for this Observation? In Order, that if there is a necessity of erecting some buildings, proper measures may be taken to send Workmen &c.; 3) That if there is not a sufficient number of astronomers belonging to the Academy, to render this observation compleat . . . I propose, and I undertake to cause search, among the Sea Officers, for proper persons, who, during the time which yet remains till the passage of Venus, may be instructed and practised for this Observation, under the care and direction of the Professors, in order to be employed in this Expedition with advantage. . . .[109]

The Academy was of course delighted to fulfill these requests, and though there is no mention of this particular aspect of Russian preparations in Catherine's letter, letters were again sent abroad inviting foreign scientists to observe the transit from Russian soil, at the expense of the Empress and in cooperation

[108] ibid., p. 378.
[109] "Letter of her Imperial Majesty of all the Russias to his Excellency the Count Wolodimer Orlow, Director of the Academy of Sciences at Petersburg," Letters and Papers of the Royal Society (1741-1806), Decade v, Vol. 41, No. 1, fol. 1. This is a copy which was sent to James Short by Rumovsky to inform him of Russian activities concerning the transit of 1769. Catherine's original letter was dated 3 March 1767, but it was communicated to the Royal Society 14 January 1768, JBRS, XXVI (1767-1790), 416-419.

with the Imperial Academy. Though that invitation met no response at the Académie des sciences, as we have already noted, other European scientists were happy to go. The German Jesuit, Christian Mayer, went to St. Petersburg; the Swiss astronomers Mallet and Pictet also responded; and the Swede who was afterward to become an associate of the Imperial Academy, Johannes Lexell, long anxious to work under the direction of the Russian Academy's great mathematician, Léonhard Euler, also took advantage of the general transit invitation to make the voyage.[110] Incidentally, Christian Mayer, Professor of Mathematics and Experimental Physics at Heidelberg, was the same astronomer whose services the Elector Palatine had offered to the Royal Society.

Besides the problem of staffing the planned expeditions, the Imperial Academy also faced the difficulty of supplying them with instruments, a difficulty which at first seemed insurmountable. But James Short in London assured Rumovsky, who had written on behalf of the Academy, that he could supply the necessary instruments. It was a substantial order, and the list of "Instruments making in London, for the use of the Imperial Academy," which Short drew up for the benefit of the Royal Society is a measure of both the dependence of the Imperial Academy on foreign manufacturers of scientific instruments and the capacity of British ateliers to produce them in the expanding, lucrative trade in such equipment in the latter half of the eighteenth century. The list included four quadrants; an astronomical clock; two achromatic telescopes with a focal length of 18 feet and 3½ feet, respectively; three Gregorian reflecting telescopes, one of 36-inch and two of 24-inch focus, and micrometer attachments for two of these; six achromatic telescopes of 12-foot focus; two of the same of 8-foot focus with micrometer attachments by Graham; and eight 3-foot achromatic telescopes with wide-field characteristics.[111] This came to twenty-one tele-

---

[110] I. Lubimenko, "Un académicien Russe à Paris (d'après ses lettres inédites, 1780-1781)," *Revue d'histoire moderne*, N.S., 20 (Novembre-Decembre 1935), 417.

[111] "Instruments making in London, for the use of the Imperial Academy of Sciences at St. Petersburg," *Letters and Papers of the Royal Society*, *op.cit.*, fol. 3.

scopes alone, an order which any modern manufacturer would cherish.

This very substantial order for instruments finds ready explanation in the actual number of Russian stations. Not counting those observations made in St. Petersburg, there were six widely scattered Russian expeditions. At an extreme east longitude of 130 degrees, there was the Yakutsk station, under a Russian officer, Captain Islenieff.[112] Two stations were set up in the southern Ural Mountain cities of Orsk and Orenburg, between east longitudes of 55 and 60 degrees; and to these, Euler and W. L. Krafft were sent.[113] It was originally planned to establish an observational post in Astrakhan, at the mouth of the Volga.[114] When the observations were actually made, however, the station was much farther east on the Caspian littoral, in Gurief (or Gurev) at the mouth of the Ural River. There the team of Lowitz and Inochodsow were sent.[115] Finally, two stations were set up on the Kola Peninsula, between the Arctic Ocean and the White Sea; and to Ponoi at its very tip went the Swiss observers, Mallet and Pictet, while to Kola, near Murmansk, went the all-Russian team of Rumovsky and Ochtenski.[116]

[112] J. Bernoulli, III, *Recueil pour les astronomes* (Berlin: n.p., 1771), I, 199. Islenieff's observations were published in Russian.

[113] *ibid.*, p. 199. These observations were published in German. See Wolfgang Ludwig Krafft, *Auszug aus den Beobachtungen welche zu Orenburg bey Gelegenheit des Durchgangs der Venus vorbey der Sonnenscheibe angestellt worden sind* (St. Petersburg: Kayser. Acad. der Wissen., 1769).

[114] "Letter from M. Rumovsky of the Imperial Academy of Sciences at Petersburg to Mr. Short of the Royal Society of London," *Letters and Papers of the Royal Society, op.cit.*, fol. 2. The letter is dated St. Petersburg, 23 October 1767.

[115] Bernoulli, *op.cit.*, p. 199. These observations were also published in German. See George Moritz Lowitz, *Auszug aus den Beobachtungen welche zu Gurief bey Gelegenheit des Durchgangs der Venus vorbey der Sonnenscheibe angestellt worden sind* (St. Petersburg: Kayser. Acad. der Wissen., 1770).

[116] Bernoulli, *op.cit.*, pp. 198-200, makes the statement that all of the Russian-sponsored observations were successively published in seven small brochures, of which those by Krafft and Lowitz are obviously two, but I have never been able to discover even the titles of the others. Mallet's observations are, however, available in "Extract of a Letter from Mr. Mallet, of Geneva, to Dr. Bevis, F.R.S.," *Phil. Trans.*, 1770, pp. 363-367.

In any case, all of them were put into an enormous Latin synthesis of the Russian observations which Christian Mayer put together under the following equally enormous title: *Ad augustissimam Russiarum omnium Imperatricem Catherina II Alexiewnam, expositio de transitu Veneris ante discum Solis de 23 Maii 1769, jussu illustrissimi & excellentissimi Domini D. Comitis Woldimeri ab Orlow illustr. Academiae Scientiarum directoris suscepta, ubi agitur de fine hujus ob-*

. With the location of the Russian observers, we have dealt with the last of the international expeditions, either overseas or across the span of a continent. The Swedes again sent observers to special stations within their national boundaries, but these need no special elaboration. As with the important observations of 1761, and for the same reasons, the best way of presenting those of 1769 will again be in a summarizing table, identical in form and notation with the one in the previous chapter. Together with those utilized in the construction of the first table, the following additional sources of data (with their abbreviations), will be cited: *Memoirs of the American Academy of Sciences* (Am.Ac.), *Transactions of the American Philosophical Society* (APS), and the *Collectio omnium observationum quae occasione Transitus Veneris per Solem A. MDCCLXIX. . . .* (Col.) of the Petersburg Academy. The volume by J. F. Encke referred to in this table is *Der Venusdurchgang von 1769 als Fortsetzung der Abhandlung über die Entfernung der Sonne von der Erde.*

OBSERVATIONS OF THE TRANSIT OF VENUS, 1769

PLACE	OBSERVERS	INSTRU-MENTS	NATIONAL SPONSOR/OR NATION-ALITY	TYPE OF OBSERVA-TION	SOURCE
1. Yakutsk	Islenieff	10'	Russia	I,2;E,1-2	Col., 329 338, 339
2. Ibid.		15'	Russia	E,1-2	Ibid.
3. Manila	de Ronas		Spain	E,1-2	Chappe, Voy. en Cal., 158
4. Peking	Dollières	18'	French Jesuit	E,1-2	Encke, 58-59

servationes, 1) cognoscendi veram parallaxin horizontalem Solis, 2) determinandi veram distantium Solis a Tellure, 3) ceterorumque planetarum & commentarum ordinem & distantium, 4) deque commodis inde natis pro geographia, re nautica, physica, &c. adductis ubique observationibus, earumque calculis ac methodus, ipsaque parallaxi hinc deducta (Patropoli: Typis Academiae Scientiarum, 1769). Much of this work is devoted to a general treatise on astronomy, the transit of Venus serving as the precipitating factor in bringing it into existence. The transit and allied observations in Russia were also separately issued as *Collectio omnium observationum quae occasione transitus Veneris per Solem A. MDCCLXIX per Imperium Russicum institutae fuerunt* (Petropoli: 1770), and also put into the *Novi Commentarii*, xiv, 2 (1769), 111-569.

PLACE	OBSERVERS	INSTRU-MENTS	NATIONAL SPONSOR/OR NATION-ALITY	TYPE OF OBSERVA-TION	SOURCE
5. Ibid.	Collas	14′	Ibid.	E,2	Ibid.
6. Batavia	Mohr	3′R	Holland	E,1-2	Ph.Tr. 1771, 435
7. Dinapur	Degloss	2′R	Britain	E,1-2	Ibid., 1770, 239
8. Phesabad	Rose		Britain	E,1-2	Ibid., 444-450
9. Orsk	Euler, C.	12′	Russia	E,1-2	Col., 277
10. Orenburg	Krafft	12′	Russia	E,1-2	Col., 225-226
11. Ponoi	Mallet	12′A	Russia	I,1-2	Col., 37
12. Gurief	Lowitz	12′	Russia	E,1-2	Col., 198
13. Ibid.	Inochodsow	2′R	Russia	E,1-2	Col., 199
14. Kola	Rumovsky	12′	Russia	I,2;E,1-2	Col., 152-153
15. Ibid.	Ochtenski	2′R	Russia	E,1	Ibid.
16. Vardö	Hell	10′A	Denmark	I,1-2;E,1-2	Newcomb, 301-305
17. Ibid.	Sajnovics	10½′	Denmark	I,1-2;E,2	Ibid.
18. Ibid.	Borgrewing	8½′	Denmark	I,2;E,2	Ibid.
19. St. Petersburg	Mayer, C.	18′	Russia	E,2	Col., 12-14
20. Ibid.	Euler	7′	Russia	E,2	Ibid.
21. Ibid.	Lexell	2½′R	Russia	E,2	Ibid.
22. Ibid.	Stahl	3½′R	Russia	E,2	Ibid.
23. Cajaneborg	Planmann	12′	Sweden	I,2;E,2	Sch.Ab., 1769, 212
24. Ibid.	Uhlwyk	3′A	Sweden	E,2	Ibid.
25. North Cape	Bayley	22′R	Britain	I,2	Ph.Tr., 1769, 266
26. Hammerfest	Dixon	2′R	Britain	I,2	Ibid., 253-262
27. Wanhalinna-berg, near Abo	Gadolin	20′	Sweden	I,1-2	Sch.Ab., 1769, 173
28. Ibid.	Justander	3′	Sweden	I,1-2	Ibid.
29. Stockholm	Wargentin	21′	Sweden	I,1-2	Ibid., 149-154
30. Ibid.	Ferner	10′	Sweden	I,1-2	Ibid.
31. Ibid.	Wilke	1½′R	Sweden	I,1-2	Ibid.

	PLACE	OBSERVERS	INSTRU-MENTS	NATIONAL SPONSOR/OR NATION-ALITY	TYPE OF OBSERVA-TION	SOURCE
32.	Hernosand	Gissler	22′	Sweden	I,1-2	Ibid., 225-226
33.	Uppsala	Prosperin	16′	Sweden	I,1-2	Ibid., 156-158
34.	Ibid.	Strömer	3′R	Sweden	I,1-2	Ibid.
35.	Ibid.	Melander	20′	Sweden	I,1-2	Ibid.
36.	Ibid.	Bergmann	21′	Sweden	I,1-2	Ibid.
37.	Ibid.	Salenius	12′	Sweden	I,1-2	Ibid.
38.	Greifswald	Mayer, A.	7′	German	I,1-2	Ph.Tr., 1769, 284
39.	Ibid.	Röhl	16′	German	I,1-2	Encke, 62-63
40.	Lund	Schenmark	21′	Sweden	I,1-2	Sch.Ab., 1769, 223
41.	Ibid.	Nenzelius	20′	Sweden	I,1-2	Ibid.
42.	Lübeck	Brashe	?	?	I,1	Encke, 62-63
43.	Ibid.	?	?	?	I,2	Ibid.
44.	Ibid.	?	?	?	I,2	Ibid.
45.	Bützow	?	?	?	I,2	Ibid.
46.	Kiel	Ackermann	4′R	?	I,2	Ibid.
47.	Paris, Col. L. le Gr.	Messier	12′A	France	I,2	Mém., 1771, 504
48.	Ibid.	Baudouin	3′A	France	I,2	Ibid.
49.	Ibid.	Turgot	11 in. R	France	I,2	Ibid.
50.	Ibid.	Zannoni	3′R	France	I,2	Ibid.
51.	Paris, Obs.Roy.	Cassini de Thury	3½′A	France	I,2	Mém., 1769, 229, 245, 529
52.	Ibid.	Duc de Chaulnes	3½′A	France	I,2	Ibid.
53.	Ibid.	du Séjour	6′A	France	I,2	Ibid.
54.	Ibid.	Maraldi	3′A	France	I,2	Ibid.
55.	Colombes	Bernoulli	2′R	German	I,2	Newcomb, 311
56.	Passy	De Bory	5′A	France	I,2	Mém., 1769, 531
57.	Ibid.	De Fouchy	2½′R	France	I,2	Ibid.
58.	Chat. de Hubert	Le Monnier	10′A	France	I,2	Ibid., 187
59.	Ibid.	Chabert	18′	France	I,2	Ibid.
60.	Saron	De Saron	3½′A	France	I,2	Ibid., 421-422

	PLACE	OBSERVERS	INSTRU- MENTS	NATIONAL SPONSOR/OR NATION- ALITY	TYPE OF OBSERVA- TION	SOURCE
61.	Kergars	d'Après	?	France	I,2	Encke, 62-63
62.	Toulouse	d'Arquier	?	France	I,2	Mém., 1769, 421-422
63.	Ibid.	Garipui	?	France	I,2	Ibid.
64.	Rouen	Dulague	?	France	I,2	Ibid.
65.	Ibid.	Bouin	?	France	I,2	Ibid.
66.	Havre de Grace	Diquemar	5′	France	I,1-2	Ibid.
67.	Bordeaux	Faugère	2½′R	France	I,2	Ibid., 510-512
68.	Ibid.	De la Roque	2′	France	I,2	Ibid.
69.	Caen, the Mission	Pigott, Sr.	6′A	Britain	I,1-2	Ph.Tr., 1770, 262
70.	Ibid.	Pigott, Jr.	1½′R	Britain	I,2	Ibid.
71.	Ibid.	Rochfort	3′A	France	I,2	Ibid.
72.	Caen, Piggot Obs.	?	17 in.	?	I,1-2	Ibid., 264
73.	Brest	Verdun	7′	France	I,2	Ibid., 346
74.	Ibid.	Fortin	5′A	France	I,2	Ibid.
75.	Ibid.	Blondeau	5′A	France	I,2	Ibid.
76.	Ibid.	Le Roy	14′	France	I,2	Ibid.
77.	Greenwich	Maskelyne	2′R	Britain	I,1-2	Newcomb, 314
78.	Ibid.	Hitchins	6′R	Britain	I,1-2	Ibid.
79.	Ibid.	Hirst	2′R	Britain	I,1-2	Ibid.
80.	Ibid.	Horsley	10′A	Britain	I,1-2	Ibid.
81.	Ibid.	Dunn	3½′A	Britain	I,1-2	Ibid.
82.	Ibid.	Dolland, P.	3½′	Britain	I,1-2	Encke, 60-61
83.	Ibid.	Nairne	2′R	Britain	I,1-2	Ibid.
84.	London, Mid. Temp.	Horsfall	R	Britain	I,1-2	Ph.Tr., 1769, 171
85.	Ibid., Spit. Sq.	Canton	?	Britain	I,1-2	Ibid., 192
86.	Ibid., Aust. Frs.	Aubert	2′R	Britain	I,1-2	Ibid., 378
87.	Windsor Castle	Harris	1½′	Britain	I,1-2	Ibid., 427
88.	Shirburn Castle	Macclesfield	3½′	Britain	I,1-2	Ibid., 173
89.	Ibid.	Lady Maccles- field	6′R	Britain	I,2	Ibid.

PLACE	OBSERVERS	INSTRU-MENTS	NATIONAL SPONSOR/OR NATION-ALITY	TYPE OF OBSERVA-TION	SOURCE
90. Ibid.	Bartlett	14'	Britain	I,1-2	Ibid.
91. Leicester	Ludlam	3'A	Britain	I,1-2	Ibid., 238
92. Oxford	Hornsby	7½'A	Britain	I,1-2	Ibid., 175-185
93. Ibid.	Lucas	6'	Britain	I,1	Ibid.
94. Ibid.	Clare	6'	Britain	I,2	Ibid.
95. Ibid.	Sykes	3½'A	Britain	I,1-2	Ibid.
96. Ibid.	Shukburgh	?	Britain	I,1-2	Ibid.
97. Ibid.	Nikitin	10 in. R	Britain	I,1-2	Ibid.
98. Ibid.	Williamson	8'	Britain	I,1-2	Ibid.
99. Ibid.	Horsley	1½'R	Britain	I,1-2	Ibid.
100. Ibid.	Jackson	9'	Britain	I,2	Ibid.
101. Kew	Bevis	3½'	Britain	I,1-2	Ibid., 189
102. Hawkhill	Alemoor	18 in. R	Britain	I,1-2	Ibid., 339
103. Ibid.	Hoy	3½'A	Britain	I,1-2	Ibid.
104. Ibid.	Lind	2'A	Britain	I,2	Ibid.
105. East Dereham	Wollaston	1'R	Britain	I,1	Ibid., 407-413
106. Kirknewton	Brice	?	Britain	I,2	Ibid., 345
107. Glasgow, near Univ.	Wilson, A.	?	Britain	I,1-2	Ibid., 333
108. Ibid.	Wilson, P.	1'R	Britain	I,1-2	Ibid.
109. Ibid.	Williamson	2½'	Britain	I,1-2	Ibid.
110. Hinkley	Robinson	?	Britain	I,1-2	Encke, 60-61
111. Leyburn	G. G.	1½'R	Britain	I,1-2	Ibid.
112. Cape Lizard	Bradley	2'	Britain	I,1-2	Ibid.
113. Cavan, Ireland	Mason	2'	Britain	I,1-2	Ph.Tr. 1770, 462, 463, 488
114. Agromonte	de Queiros	?	Spain	I,2	Encke, 62-63
115. Cadiz	Tofino	7'	Spain	I,1-2	Ibid.
116. Gibraltar	Jardine	2'R	Britain	I,1-2	Ph.Tr., 1769, 347
117. Isle Coudre	Wright	2'R	Britain	I,1-2	Ibid., 276
118. Newbury, Mass.	Williams	R	Britain	I,1-2	Am. Ac., I, 111
119. Cambridge, Mass.	Winthrop	2'R	Britain	I,1-2	Ph.Tr. 1769, 356
120. Providence, R.I.	West	3'R	Britain	I,1-2	APS, I, 95-96

	PLACE	OBSERVERS	INSTRU-MENTS	NATIONAL SPONSOR/OR NATION-ALITY	TYPE OF OBSERVA-TION	SOURCE
121.	Cape Francis, S.D.	Fleurieu	2½′A	France	I,1-2	Mém., 1769, 516
122.	Ibid.	La Filière	3′	France	I,1-2	Ibid.
123.	Ibid.	Destourès	2′	France	I,1-2	Ibid.
124.	Ibid.	Pingré	5′A	France	I,1-2	Ibid.
125.	Martinique	Christophe	?	France	I,1-2	Encke, 64-65
126.	Baskenridge, N.J.	Alexander	?	Britain	I,1-2	APS, I, 125
127.	Lewes, Del.	Biddle	R	Britain	I,1-2	Ibid., 89
128.	Ibid.	Bayley	4½′	Britain	I,1-2	Ibid.
129.	Philadelphia	Shippen	2′R	Britain	I,1-2	Ibid., 45
130.	Ibid.	Williamson	24′	Britain	I,1-2	Ibid.
131.	Ibid.	Thomson	1′R	Britain	I,2	Ibid.
132.	Ibid.	Prior	1½′R	Britain	I,1-2	Ibid.
133.	Ibid.	Ewing	4½′R	Britain	I,1-2	Ibid.
134.	Ibid.	Pearson	1′R	Britain	I,1	Ibid.
135.	Norristown	Lukens	42′	Britain	I,1-2	Ibid., 24-29
136.	Ibid.	Rittenhouse	36′	Britain	I,1-2	Ibid.
137.	Ibid.	Smith	2′R	Britain	I,1-2	Ibid.
138.	Wilmington, Del.	Poole	12′	Britain	I,1-2	Ibid., 126
139.	Talbot Cnty., Md.	Leeds	2′R	Britain	I,1-2	Ph.Tr., 1769, 444
140.	Quebec	Holland	?	Britain	I,1	Ibid., 247-252
141.	Hudson's Bay	Dymond	2′R	Britain	I,2;E,1-2	Ibid., 480
142.	Ibid.	Wales	2′R	Britain	I,2;E,1-2	Ibid.
143.	Mexico	Alzate	?	Spain	I,2	Encke, 64-65
144.	San José, Calif.	Chappe	3′A	France	I,2;E,1-2	Chappe, Voy. en Calif., 94
145.	Ibid.	Doz	?	Spain	I,2;E,1-2	Ibid., 159
146.	Ibid.	Medina	?	Spain	I,2;E,1-2	Ibid.
147.	St. Anne, Calif.	Velasquez	?	Spain	I,2;E,1-2	Ibid.
148.	Tahiti	Green	2′R	Britain	I,1-2;E,1-2	Ph.Tr., 1771, 410
149.	Ibid.	Cook	2′R	Britain	I,1-2;E,1-2	Ibid.
150.	Ibid.	Solander	3′R	Britain	I,1-2;E,1-2	Ibid.

Once again, without exaggerating its importance, a statistical analysis of this table reveals some interesting aspects of the 1769 observations and the changes which have taken place in comparison with those of 1761. From a third place position with regard to the first transit (with 18 significant observations), the British have moved, at least in their quantitative interest in the problem, to first place with 69 separate observations.[117] And even though the French have increased the number of their transit observations, they are now in the second position with less than half (34) the number of British observations. It would be folly to say, on this basis and with the benefit of long historical perspective, that this is symbolic of the evolving pattern of the history of science in this period, an early warning of the future decline of French scientific hegemony in Europe, yet the suggestion is there and the alteration in the decade quite startling. Not much can be said of the remaining national groupings or sponsors. The Italians have disappeared from the scene, save as individuals working under others, as for example, those present at the Paris Observatory. Both Spain and Russia have increased their astronomical activity, but even though there has been a fourfold expansion with respect to the latter (from three to thirteen observers) with a growth in the number of native Russian observations, the quantities in both cases are too small to tell us anything about trends. For that matter, and this further reveals the weakness in simply reading statistics, official Russian science (that is, the direction and operation of the Imperial Academy) is, in spite of the rise to prominence of men like Rumovsky and Lomonosow[118] during this period, still in the

[117] George III even had a special observatory built at Kew (from which the Bevis observation in the table was not made), so that he could personally observe the 1769 transit. A manuscript notebook kept by Dr. Demainbray entitled *Observations on Transit of Venus, 1769, Richmond Observatory*, is preserved in the Library of King's College, London, but I have not seen it. From a letter by J. Chaldecott, Dept. of Physics, Science Museum, South Kensington, London, to Professor Henry Guerlac, Cornell University, dated 12 November 1953.

[118] On Lomonosow as astronomer, see Otto Struve's excellent little article, "Lomonosow," *Sky and Telescope*, xiii, 4 (February 1954) 118-120. Lomonosow died in 1765, so he played no part in the Russian observations of the second transit. But he did observe the first, from his house in Saint Petersburg with a 4½-foot refractor. He "felt curious about the physical aspects of the phenomenon," and together with A. D. Krassilnikov and N. G. Kourganov, who had observed the

hands of distinguished foreigners. Swedish activity in 1769 remains at about the same level at which it had been in 1761, with highly competent astronomers observing from important northern stations.

Another interesting truth inherent in the structure of the table concerns the instruments. In 1761, the majority of the telescopes used were nonachromatic refractors, numbering 66, followed by 40 reflectors and 3 achromatic instruments. By 1769, there were important changes, indicative of the improved manufacture of instruments, which in turn was partly brought about by the large demand for high quality equipment created by events such as the transits. The number of reflectors in use for the second transit rose to 49, actually not too great a change considering the increased number of observers. But there was a sharp drop in the use of nonachromatic instruments in favor of the achromatic lens systems, which rose to 27, a ninefold increase, most of which derived from the English workshops. It follows, too, that English observers were therefore likely to be better equipped, but it is dangerous to make such generalizations, for, as we have already seen with regard to the Russian acquisition of instruments, a great deal of English equipment was exported to foreign buyers.

What then of the results of 1769 for the contemporary astronomical world. Was the eighteenth-century dream of realizing the actual compass of its physical world to be achieved? Did the increased effort of 1769, numbering 151 observers from 77 stations produce a definitive solar parallax? Sadly, the answer is "no" for the century of the Enlightenment, though the nineteenth-century reconsiderations of its predecessor's work were to narrow considerably the dimensions of uncertainty behind that "no." The number of papers on the value of the solar parallax deducible from the 1769 transit was enormous; about two hundred were sent to the Académie des sciences, and prob-

---

transit from Delisle's old observatory, wrote "The Appearance of Venus on the Sun as Observed at the St. Petersburg Academy of Sciences on May 26, 1761," which was not published during his lifetime. It seems to have been missed also by historians of science, including those interested in the transits. Toward the end of the nineteenth century, it was printed in the fifth volume of M. I. Sukhomlinov's edition of Lomonosow's collected works, *ibid.*, p. 118.

ably as many as four hundred more to the remaining world-wide scientific bodies.[119] But only a very few of these are worth considering.

One of the most uncertain calculations of all was William Smith's of Philadelphia, based as it was on a comparison of the very indecisive contacts of ingress between those recorded at Norristown, Pennsylvania, Greenwich, and other European stations. Smith even made use of the first outer contact between Venus and the sun when the planet first begins to dent the solar disk. Generally speaking, this is the least useful observation of all, so that his conclusion of 8.6045 seconds as the mean solar parallax carried no assurance of certainty with it.[120] As he had done with the first transit, Thomas Hornsby again calculated the solar parallax. His work was based on the method of durations, derived from a knowledge of the times of inner contacts. But his value of 8.78 seconds was also based on special conditions, such as the exclusion of the Cajaneborg data and the use of contact values for Tahiti based on averages of the separate observations there.[121]

Of the French evaluations of the parallax, Pingré's and Lalande's are easily the most important. Pingré examined the problem twice, in two memoirs published in the volumes of the Académie for 1770 and 1772. In the 1770 paper, he first compared the Hudson's Bay observations with those of Vardö and a combination of ingress observations at European stations and egress observations at St. Petersburg. This very unsatisfactory technique gave him a parallax of 9.2 seconds, but as soon as the California and Tahiti data reached him, this result was modified to 8.88 seconds.[122] Two years later, as a result of the

[119] R. A. Proctor, *Transits of Venus* (New York: R. Worthington & Co., 1875), p. 85. Proctor cites Dubois, *Les passages de Vénus* . . . as the source of the first figure, and simply presents the second estimate as his own. Though I have mentioned it on their authority, the quantity seems greatly exaggerated.

[120] W. Smith, "The Sun's Parallax deduced from a Comparison of the Norriton Observations of the Transit of Venus, 1769; with the Greenwich and other European Observations of the same," *Trans. Am. Phil. Soc.*, 1, 174. All references to these Transactions are to the second edition of 1789.

[121] T. Hornsby, "The Quantity of the Sun's Parallax, as deduced from the Observations of the Transit of Venus, on June 3, 1769," *Phil. Trans.*, 1771, p. 579.

[122] Pingré, "Examen critique des observations du passage de Vénus sur le disque du Soleil, le 3 Juin 1769; et des conséquences qu'on peut légitimement en tirer," *Mémoires*, 1770, p. 583.

influence of Hornsby's work, Pingré altered the latter calculation, to obtain 8.80 seconds as his final figure.[123] Lalande also published his results in two papers,[124] obtaining one of the smallest of all the eighteenth-century calculations, 8.50 seconds, as reported by Encke and Procter, but in his own words, the mean solar parallax "est renfermé dans les limites de 8″, 55 à 8″, 63. . . ."[125]

Of the other contemporary evaluations, those of Planmann, Hell, and Lexell need only be mentioned as examples of the variants introduced from the Swedish, Russian, and Viennese corners of the astronomical world. Planmann produced the smallest value of all to emerge from the 1769 transit, 8.43 seconds, apparently because he did not minimize his own observations at Cajaneborg as others had done.[126] Hell gave the solar parallax a value of 8.70 seconds in successive editions of his *Ephemerides Vindobonenses* for 1773 and 1774, but apparently satisfying no one but himself as to the soundness of his method.[127] Finally, Lexell, basing his work on Léonhard Euler's tremendous analytical treatise on the solar parallax,[128] sought to perform the definitive calculations for its evaluation. In the first of his two papers, published in 1771, he obtained a parallax of 8.68 seconds, but in 1772, when he presented the second paper to the public, this was reduced to 8.63 seconds.[129]

---

[123] Pingré, "Mémoire sur la parallaxe du Soleil, déduite des meilleures observations de la durée du passage de Vénus sur son disque le 3 Juin 1769," *Mémoires*, 1772, I, 419. In this paper Pingré considered the methods and results of Euler, Hell, and Lalande, as well as those of Hornsby.

[124] Lalande, "Mémoire sur la parallaxe du Soleil, qui résulte du passage de Vénus, observée en 1769," *Mémoires*, 1770, 9-14, and "Mémoire sur la parallaxe du Soleil, déduite des observations faites dans la mer du Sud, dans le royaume d'Astracan, & à la Chine," *ibid.*, 1771, pp. 776-799.

[125] Lalande, "Mémoire sur la parallaxe du Soleil . . . ," *ibid.*, 1771, p. 798.

[126] Quoted by Encke, *Der Venusdurchgang von 1769* . . . , p. 22, from the *Schwedische Abhandlungen*, XXXIV (1772), 179.

[127] The opinion is Encke's, *op.cit.*, pp. 24-25.

[128] L. Euler, "Expositio methodorum cum pro determinanda parallaxi Solis, ex observato transitu Veneris per Solem, tum pro inueniendis longitudinibus locorum super terra ex observationibus eclipsium Solis, una cum calculis et conclusionibus inde deductis," *Novi Commentarii*, XIV, 2 (1769), 321-554.

[129] J. Lexell, "De parallaxi Solis conclusa ex transitu Veneris per Solem A. 1769 in insula Regis Georgii [Tahiti] observato," *ibid.*, XVI (1771), 586-648, and "Disquisitio de investiganda parallaxi Solis, ex transitu Veneris per Solem anno 1769," *ibid.*, XVII (1772), 609-672.

Thus the transit of 1769 produced its own range of values for the solar parallax, from Planmann's 8.43 seconds to Pingré's 8.80 seconds. This was a considerable advance over the situation following the first transit, which had produced limits of 8.28 and 10.60 seconds. In effect, the new results created a permissible range of variation of 0.37 seconds in which the estimate of parallax could oscillate above 8.43 and below 8.80, as against a value of 2.32 seconds for the calculations after 1761. But though this was a moderately satisfactory conclusion, the mere fact that different values for the solar parallax could be obtained by giving more or less weight to certain observations meant that the established figures would vary under every later consideration.

Indeed, such reconsideration was given to the transits of Venus many times during the nineteenth century, but the two most important re-evaluations of this method of determining the solar parallax were certainly those put forth by J. F. Encke in 1824 and S. Newcomb in 1890. Encke's analysis of the 1769 transit gave him a parallax of 8.6030 seconds, which he combined with his evaluation of the first transit to obtain a combined figure of 8.5776 as the definitive solar parallax.[130] This value was upheld in most astronomical papers for about twenty-five years. But about the middle of the century, other techniques of investigating the problem were developed by Hansen, who used the motion of the moon as the basis for his study, and Le Verrier, who used both planetary perturbations and the motion of the sun and moon around their common center of gravity as his foundation, to name only two of the outstanding men in this field. Their work led to the conclusion that Encke had overestimated the solar distance.[131]

The doubt thus cast upon Encke's work also tended to reflect adversely upon the transits of Venus as a whole. It was to restore the transit technique that other astronomers like Powalky and Stone addressed themselves to the eighteenth-century data, to arrive at values for the solar distance which re-

[130] Encke, *Der Venusdurchgang von 1769* . . . , p. 108.
[131] The relationship between solar parallax and solar distance, in terms of mileage, can be quickly grasped from the table in Appendix II.

duced Encke's figure considerably. That story, however, is really beyond the scope of this study.[132] Even though Newcomb discussed the problem after the nineteenth-century transits of Venus, he limited the bulk of his efforts to the data of the eighteenth, thereby continuing the process of sustaining the technique as a whole. He deviated from others who had utilized the original material by making use of the improved longitudinal calculations of his own day wherever possible. His solar parallax of 8.79 seconds[133] was therefore one of the best to derive from the eighteenth-century transits, comparing favorably with the very sophisticated calculation by W. Harkness, of 8.809 seconds.[134] Incidentally, this latter figure is equivalent to a solar distance of 92,796,950 miles in which there is a possible error of plus or minus 60,000 miles. A recent estimate of the solar distance by H. Spencer Jones gives it as 93,005,000 miles with an uncertainty, either way, not exceeding 9,000 miles.[135]

But the key problem in 1769 was once again the determination of the moments of contact between the planet and the sun. Experience with the black-drop effect in 1761 had indeed alerted astronomers to the idea of real and apparent contact, with the former defined as the moment when the elongating matter of the planet appears to break, and the latter when the black-drop first forms. But the attempt to turn this undesirable phenomenon to advantage was without success, for it was much too variable. For example, at Hudson's Bay, Wales found a 24 second difference between real and apparent contact at egress; at Tahiti, Green found it to be 48 seconds, and Cook recorded 32 seconds; while at Greenwich, the values ranged from 26 to 52 seconds.[136] And this sort of notice could be extended across the entire span of observations in 1769.

Thus, for the eighteenth-century transits as a whole, the problem of longitudinal precision and an inability to handle the

[132] See Proctor, op.cit., pp. 89-92 for a fair evaluation of their work.

[133] Newcomb, op.cit., p. 402.

[134] W. Harkness, "The Solar Parallax and its Related Constants Including the Figure and Density of the Earth," Washington Observations (1885), Appendix III, 140.

[135] H. Spencer Jones, "The Distance of the Sun," Endeavour, I, 1 (January 1942), 17.

[136] Proctor, op.cit., p. 83.

black-drop effect were major factors in upsetting the transit method in the eyes of contemporary astronomy. But there were other factors. Careful analysis of the problem within the framework of the eighteenth-century scientific structure reveals that there were at least three important difficulties which had not been considered seriously enough at the outset: the amount of variation introduced by the personality and physiology of the individual observer, the flaws in the telescopes used, and the conditions of vision. The development of error-theory would one day help to adjust scientific data for some of these factors, but this technique was not yet a part of eighteenth-century science.

The variation introduced by the first factor sometimes amounted to nearly a minute in observations that were expected to be correct within a second.[137] That this may not have always been due to the phenomenon itself can be surmised from an incident in connection with David Rittenhouse. For there is good reason to believe that even so careful an astronomer could fall from scientific grace. He was so overcome by excitement at one of the key moments of observation, that his measure of time was definitely distorted.[138]

The most serious flaw introduced by the telescopes used was an irradiation effect inherent in their structures. This varied from one instrument to another according to its focal length and type. Three kinds of telescopes were used, as we have already had occasion to mention: reflecting instruments ranging in size from 1 to 7 feet, achromatic refractors from 2 to 10 feet, and nonachromatic refractors from 2½ to 42 feet.

Instrument irradiation produces a scattering of light that creates a fuzzy image, a result similar to the effect brought about by the atmospheric conditions of vision. As a matter of fact, the gradation of imperfections in vision due to instruments varied through a range of effects comparable to those produced by the atmosphere alone between zenith and an altitude of three degrees.[139] In other words, the total scattering or irradia-

[137] *Trans. Am. Phil. Soc.*, 1 (1771), 26.
[138] Newcomb, *op.cit.*, p. 323.
[139] *ibid.*, p. 380.

De par Le Roy

A tous gouverneurs et nos Lieutenans generaux en nos provinces et armées, gouverneurs particuliers et Commandans de nos Villes, places et troupes, Et à tous autres nos officiers, Justiciers et sujets qu'il appartiendra, Salut. Nous voulons et vous mandons très expressement que vous ayez à laisser seurement & librement passer les Pingré & L'Academie des Sciences allant aux Inde orientales

sans lui donner ny souffrir qu'il lui soit donné aucun empêchement, Mais au contraire tout l'ayde et assistance dont il aura besoin: le present passeport valable pour trois ans seulement: Car tel est notre plaisir. Donné à Versailles le huit Octobre 1760.

Louis

Par Le Roy

Gratis

13. Pingré's *laissez-passer*, 1761

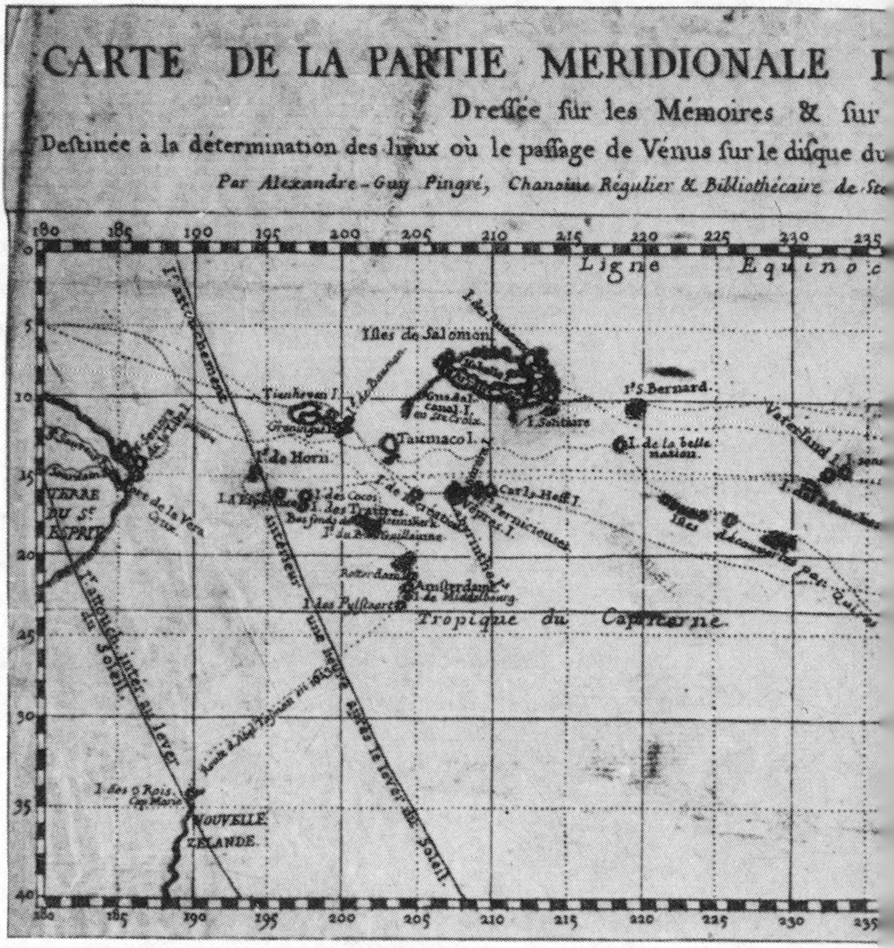

14. Carte de la mer Pacifique

MER PACIFIQUE OU MER DU SUD,

... des plus célebres Voÿageurs,

... arriver le 3 Juin 1769, peut être obſervé avec le plus grand avantage:

... nome Géographe de la Marine, de l'Académie Royale des Sciences.

Maſoce I.ᵉ
Route de Mindaña en 1605.

Route de Mindaña en 1568.

I. S. Paul
Roum de le Maire en 1616.

Route de Quiros en 1606.

Soleil

Coucher du Soleil avant dernier retranchement indirectement avant

Dernier retranchement avant coucher du Soleil

I. de Pâques.   Roum de Roggewein en 1722.

Trapiche
I.

Terre que David,
Anglois a vue, ou
a cru voir en
1686.

I. S. Ambroiſe I. de S.ᵗ Felix.

I.ᵉ Gallapagos

Guayaquil.
Cap Bl.
Paita.
PEROU
Truxillo.
le Callao de LIMA
Piſco.

Valpa-
raiſo.
I.ᵉ de Juan Fernandez.
La Conception.
I. Sta Marie.
I. Mocha.
Baldivia.

250 255 260 265 270 275 280 285 290 295 300 305

290 255 260 265 270 275 280 285 290 295 300 305

On désiroit depuis longtemps que le Passage de
Vénus sur le Soleil qui doit arriver le 6 Juin 1761
fut observé avec toute l'exactitude possible et dans les
Circonstances les plus avantageuses. L'Acad. Imperiale
de Petersbourg ayant invité l'Acad. d'envoyer
un de ses astronomes pour faire cette opération en
Sibérie, l'Académie a cru en acceptant cette
Invitation devoir envoyer de son côté aux environs
de la Pointe méridionale de l'Afrique elle a nommé
pour faire cette dernière Observation M. Pingré l'un
de ses astronomes et luy a adjoint M. Thuillier
en foy de quoy elle m'a ordonné de luy en délivrer
le présent acte qui a été signé de Messieurs les
officiers et auquel j'ay apposé le sceau de l'Académie.

Buffon

GrandJean De Fouchy

15. Commission of Pingré and Thuillier, 1761

tion of light depended on the depth of the atmosphere through which observations were made as well as on the dispersive characteristics of individual telescopes. Since most transit observations were made with the sun at low altitudes, this amounted to a considerable degree of distortion in the image viewed.

On the whole, the black-drop effect introduced the largest element of uncertainty into the method of determining the solar parallax by means of the transit of Venus. As such, it was the factor above all others that made the eighteenth-century effort to calculate the solar distance fall far short of the precise and definitive goal it had felt itself capable of reaching. Other methods, already suggested, were eventually perfected to determine the solar parallax; and other scientists were to indicate how complex a problem it really was. In the last years of the nineteenth century, for example, it was brilliantly demonstrated that the solar parallax was not an independent constant, but one intimately related to the constants of precession and nutation, the lunar parallax, the masses of the earth and moon, and the constant of aberration, to name but a few, and that a proper calculation of the solar parallax should consider all these quantities simultaneously. [140]

VI

Concern for the exact solar distance was but one element in the general attempt of eighteenth-century astronomers to give their subject the same precision which the other physical sciences were rapidly acquiring. The production of precision instruments on a large scale and their application to the problems of science in and out of the laboratory was a universal phenomenon in the eighteenth century. Apparatus which had once been designed to help understand science or to demonstrate its principle in the salon or lecture hall evolved into instruments of quantitative research in the second half of the century. The use of thermometers, barometers, and hygrometers easily demonstrate this point. Indeed, one of the best illustrations of all is the impetus which Lavoisier gave to quantitative work in chemistry by his brilliant use of the precision balance.

[140] Harkness, *op.cit.*, p. 1.

In the same way, observational astronomy during the greater part of the eighteenth century was concerned with the question of precision. Progressive increases in accuracy were obtained by refinements in the design and construction of astronomical instruments, particularly in the application of telescopic sights to older instruments, the development of lens manufacture and the use of micrometer attachments. Although this period also witnessed the full development of gravitational astronomy in the work of Euler (1707-1783), Clairaut (1713-1765), D'Alembert (1717-1783), Lagrange (1736-1813) and Laplace (1749-1827), this advance was not entirely independent of the progress in observational astronomy. This is at least true insofar as improvements in the precision to be expected from the observatory placed limits upon theoretical speculation.

At the beginning of the eighteenth century, the accuracy of astronomical observations lagged behind the achievements of theoretical astronomy. But as the century progressed the manufacture of scientific instruments fell into professional hands, and instruments of a precision hitherto unknown became a reality. Largely because of accomplishments of this sort, observational astronomers were able to undertake the task of checking theory and maintaining that dynamic balance between theory and practice which is the foundation of scientific progress. Thus it became worth while to go to different parts of the globe to measure an arc of the meridian, to determine the lunar parallax, or to deduce the shape and size of the earth. It was equally worth while to observe the transits of Venus, and to deduce the solar parallax from the collection of these observations. Not only should these investigations disclose the true dimensions of the planetary orbits, but also they should bring about a reduction of the elements of uncertainty in all calculations involving the astronomical unit.

Those who labored toward this end in the eighteenth century may be granted a true measure of success, for they succeeded in reducing considerably the range of variation in the value of the solar parallax, bringing it very close to its contemporary value. Simon Newcomb's evaluation of 8.79, on the basis of eighteenth-century data, is not too far removed from the most recent (1950) figure of 8.798,[141] and the loosely accepted solar distance of 93,000,000 miles in the present day derives almost as easily from the one parallax value as the other. Further refinements in this and other astronomical constants would, of course, continue to be made for the specialized use of the astronomer, but for all ordinary purposes the dimensions of the solar system were established by the achievement of eighteenth-century astronomy in availing itself of the opportune transits of Venus.

If the greatest hope of the "transit astronomers" was to be able to fix the "frame of the world," this was not the only significant result of their ambition. The cause of natural history was considerably advanced when the transit of Venus expeditions were linked with those engaged in the detailed exploration of the world. And while it is difficult to assess this factor quantitatively, the multiple results of this part of the transit story increased the prestige of scientific societies everywhere and lent additional weight to their future claims for a larger share of national wealth and attention. The prolonged liaison between science and government, especially with regard to the growing financial needs of science for large-scale enterprise, revealed the interdependence of science and society in very specific terms. In Russia, for example, transit activities accelerated the general growth of science that had received its initial impulse from Peter the Great and the policy of importing French and German scientists into Russia. Finally, the problem of the transits of Venus produced an intensity and breadth of effort on the part of eighteenth-century scientists that was unmatched by any other single problem. It brought to a common focus men of almost every national background with an abiding concern for the advancement of knowledge. In doing so, it helped to shape the growing international community of science and to demonstrate with striking clarity what cooperation and good will might achieve in the peaceful pursuit of truth.

[141] G. P. Kuiper (ed.), *The Sun* (Chicago: The University of Chicago Press, 1953), p. 18. The figure as actually given by Leo Goldberg, who wrote the "Introduction," is 8.79835±0.00039, or about 92,915,530 miles.

APPENDICES

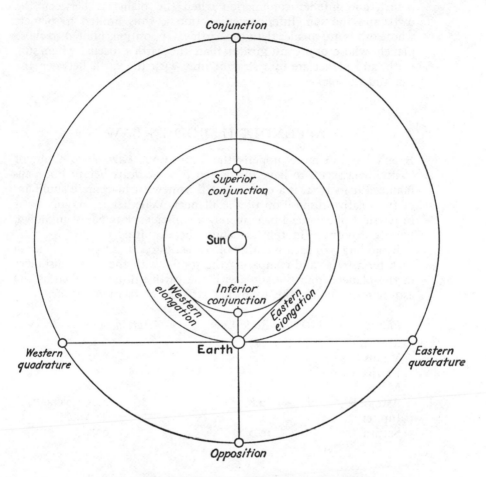

4.   PLANETARY CONFIGURATIONS

THE above diagram is designed to represent the various positions of the planets, projected on to the solar plane, and the terms used to identify those positions. A planet is said to be in quadrature when its direction, as observed from the earth, is at 90 degrees with respect to the direction of the sun from the earth. Thus, it is possible for only superior planets to be in quadrature. The angular separation of a planet from the sun is defined as its elongation, which, like quadrature, can be east or west.

Conjunction occurs when the direction of the earth, sun, and a planet coincide. They are then exactly in line, with superior conjunction taking place when the sun is between the earth and the planet, and inferior conjunction when the planet is between the earth and the sun. Inferior conjunction is thus limited to planets whose orbits are smaller than the earth's. Opposition, limited to those planets whose orbits are greater than the earth's, occurs when sun, earth, and planet are in a straight line, with the earth between the sun and the planet.

APPENDIX II. BODE'S LAW

BODE'S LAW, or more properly the Titius-Bode Law, since Titius of Wittenberg seems to have discovered it some years before Bode announced it in 1772, is a convenient if somewhat inaccurate summary of the relative disposition of the planets. Whether it has any basis in physical phenomena or is merely a coincidence remains unknown, but its expository usefulness cannot be dismissed.

By adding 4 to each member of the series 0,3,6,12,24 . . . , dividing each term by 10 and comparing the results with the actual distances of the planets from the sun (with the earth's distance taken as the astronomical unit), the following striking correlations occur:

Planet	Titius-Bode Law	Actual Solar Distance
Mercury	0.4	.39
Venus	0.7	.72
Earth	1.0	1.0
Mars	1.6	1.52
Asteroids	2.8	2.65 (average)
Jupiter	5.2	5.20
Saturn	10.0	9.54

APPENDIX III. DOCUMENTS CONCERNING ALEXANDRE-GUI PINGRÉ

[A] Lettre à Monsieurs du Fresne Marion, Capitaine & Autres Officiers de l'Etat major du Vaisseau le Comte d'Argenson

Messieurs

Nous soussignés, envoyés par le Roi & l'Académie royale des sciences, à l'isle de Rodrigue, pour y faire des observations de Mathématique & de Physique; instruits des ordres qu'il a plu au Roi de faire donner à ce sujet par le Contrôleur général de ses finances; persuadés que ces ordres vous sont parvenus par le canal de MM.

les Syndics et Directeurs de la Compagnie des Indes; informés même par les témoignages réitérés de plusieurs de MM les directeurs de la Compagnie que le plan de la Compagnie étoit que nous fussions remis en l'isle de Rodrigue, avant votre arrivée en celle de France, plan que vous nous avez toujours paru disposés à suivre jusqu'à la rencontre du vaisseau le Lys: nous ne pouvons nous permettre d'être tranquilles au sujet des bruits qui se sont répondus depuis 2 jours dans le Vaisseau. Vous êtes, dit-on, résolus à changer le plan de votre route, et à nous mener directement à l'isle de France, d'où il est très probable que nous ne pourrions être transportés à temps, c'est-à-dire avant le fin de Mai à celle de Rodrigue. D'abord, nous n'ajoutâmes aucune foi à ces bruits: une réflexion juste autorisoit certainement notre incrédulité: nous ne pouvions nous regarder comme assez étrangers dans cette affaire, encore moins comme assez au dessous de tout égard, pour qu'on ne nous est donné aucune connoissance, du moins générale, des raisons qui pouvoient autoriser de nouveaux arrangements, contraires à ceux qui avoit été précédemment déterminés. Cependant ces bruits s'acoréditent [sic]; et nous ne pouvons plus nous cacher qu'ils sont appuyés sur quelque fondement de vraisemblance. En gardant un plus long silence, nous croirions manquer aux plus saints de nos devoirs, envers le Roi, qui nous a honorés de ses ordres, et qui de sa bouche sacrée a daigné témoigner à l'un de nous deux la part qu'il prenoit à la réussite de nos observations; envers l'Académie, qui nous a accordé sa confiance, au sujet d'une observation qu'elle juge aussi importante que délicate; envers la patrie, dont nous ne craignons pas d'avancer que la gloire est intéressée au succès de notre voyage; envers l'Europe entiére, qui prévenue de notre mission, a les yeux attentifs sur la manière dont nous nous acquitterons de son objet. Nous ne répondons pas sans doute des impossibilités; mais il faut qu'elles soient réelles, & notre devoir est de les constater, autant qu'il est en nous, pour être à couvert de toute reproche.

A ces Causes, nous vous sommons, Messieurs, par le présent acte duquel nous conservons copie, pour en faire tel usage, que nous attiserons bon être, de suivre le plan de route, que vous vous étiez proposés jusqu'à la rencontre du vaisseau, le Lys; afin que nous puissions débarquer à temps à l'isle de Rodrigue. Nous protestons formellement contre tous les arrangements que vous pourriez prendre, et contre l'exécution de ceux que vous auriez déjà pris avec les Officiers du Lys, entant que ces arrangements, fondées peut-être d'ailleurs sur des faits hazardés, et sur des conjectures hors de vraisemblance, seroient contraires aux intérêts de notre commission, et aux ordres respectables, dont nous sommes vous et nous respectivement chargés. Enfin nous nous croyons autorisés à vous demander une réponse claire, catégorique et par écrit à notre sommation présente, ou en cas de refus, acte de notre protestation et de votre refus.

Fait à bord du vaisseau le Comte d'Argenson le 13e du mois d'Avril 1761. Signé Al. G. Pingré, Ch[anoine] r[égulier] de l'Académie royale des Sciences.   Thuillier.

Source: Bibl. Ste. Gen., Pingré MSS, 1803, *Relation de mon Voyage de Paris à l'isle Rodrigue*, fol. 45.

[B] Réponse à la premiere sommation présentée à M du Fresne Marion, et à MM les Officiers le 13 Avril 1761 par M Al. G. Pingré Ch.r. de l'Académie royale des sciences, et Thuilliers.

Si Messieurs les Académiciens, dénommés cy-dessus, avoient voulu observer que nous entretenons la latitude de 35 à 36 degrés, et que je m'estime ce jour par la longitude de 37 à 38 degrés, suivant le méridien de Paris, ils se seroient épargné la peine de me sommer de les conduire directement à Rodrigue; je crois être jusqu'à présent dans la route ordinaire.

Je suis les ordres que j'ai reçus de MM. les Sindics et directeurs de la Compagnie des Indes; je m'y conformerai toujours autant qu'il me sera possible, sauf à moi à en rendre compte à M. le Gouverneur de l'isle de France et à Messieurs du Conseil.

J'ajouterai en outre que le hazard m'ayant fait rencontrer le Lys, Vaisseau de la Compagnie, commandé par M. Blain des Cormiers, auquel je me reconnois subordonné, je lui ai fait part de mes ordres concernant lesdits Sieurs: en conséquence, MM les Académiciens peuvent s'adresser audit Sieur Commandant.

Comme ladite sommation des Sieurs Pingré et Thuilliers, l'un de l'Académie des sciences reconnu, et l'autre ce qu'il plaira à Dieu, ont sommé Mrs. les Officiers de mon Etat major de répondre à la dite sommation, je déclare la leur avoir communiquée, ainsi que la réponse que j'y fais, afin qu'ils puissent répondre ce qu'ils jugeront à propos.

Fait à bord du Vaisseau le Comte d'argenson, le 13 du mois d'Avril mil sept cent soixante et un. Signé Dufresne Marion.

Source: *ibid.*, fol. 45 verso.

[C]   isle de France 15 7bre 1761
A
My Lords
Les Commissaires préposés pour l'exécution de
L'Office du Lord Haut-Amiral de la grande Bretagne
et de l'Irlande &c.
My Lords,
Les plaintes que je me crois autorisés à porter devant votre tribunal contre le Sieur Robert Fletcher ne consisteront qu'en simple exposé des faits qui y donnent lieu.

Vers le fin du mois d'Aoust 1760, je fus nommé par l'Académie des sciences de Paris, pour observer en l'isle de Rodrigue ou de Diego Ruyz le *passage de Venus* devant le disque du Soleil. Une telle mission vous parut, my Lords, digne d'être secondée de vous. Pour contribuer à son succès, autant qu'il étoit en vous, vous me fites expédier en Date du 30 de Nov. de la meme année le passeport le plus gracieux, qui put vous être dicté par votre zèle éclairé pour le progrès des sciences. And whereas, y disiez vous, it is necessary that the said Monsieur Pingré should not meet *with any interruption, either in his passage to or from that Island*; you are hereby most strictly required and directed not to molest his person or effects upon any account, but to suffer him to proceed, without delay or interruption, in the execution of his design, &c. Fondé sur de tels auspices, je suis parti de France avec confiance, et je suis arrivé ici en l'isle de France ou Maurice le 5 de Mai dernier au soir. Dès le lendemain Monsieur Des Forges Gouverneur de l'isle de France ordonna d'équiper la Corvette la Mignonne pour me mettre en état d'exécuter l'objet de ma mission. Que la Mignonne n'ait été dépêchée uniquement pour me conduire à Rodrigue et me ramener ici: c'est une vérité dont le Sieur Robert Fletcher pouvoit se convaincre par la simple lecture des instructions que Monsieur des Forges avoit données au Capitaine de la Mignonne. Nous mouillâmes à Rodrigues le 28 de Mai. Il étoit éssentiel au succès de mon observation que je connuse la latitude et la longitude de lieu où elle étoit faite. Ce dernier objet nous retint à Rodrigue jusqu'à la fin de Juin, nous vîmes paroitre un navire sous pavillon Hollondois. Sa construction particuliere fit soubçonner que ce pouvoit être quelque pirate d'Angria. La Mignonne étoit en rade. Son Capitaine fit tirer sur le navire inconnu, lequel sans changer de pavillon, approcha jusqu'à la portée du pistolet arbora pavillon Anglois, et fit feu sur la Mignonne. Notre Capitaine fit aussitôt cesser le sien, donna ordre de mettre sur l'avant un pavillon Anglois, et fit crier avec le portevoix qu'il étoit vaisseau de transport. Mais le feu ennemi continua, jusqu'à ce que notre Capitaine eut fait amener son pavillon. Le Sieur Robert Fletcher soit-disant Capitaine des Vaisseaux de la Compagnie, et Commandant sur la Calapate ou le Plassei (car son navire avoit les deux noms) jugea la prise bonne parce qu'elle étoit à sa bienséance. Une autre Corvette étoit dans le port; elle se nommoit l'Oiseau; elle y avoit été envoyée pour faire une cargaison de tortues, et elle se trouvoit échouée. Le Sieur Fletcher exigea des Officiers de la Mignonne qu'ils le conduisirent près de l'Oiseau, les menaçant de les faire prendre en cas de refus. Ils refuserent, et cependant ils ne furent pas pendus, mais enfermés dans la Calapate. Le Sieur Fletcher étant descendu à terre, où il sçavoit qu'on ne pouvoit faire une grande resistance, je réclamai la Mignonne, et lui montrai mon passeport. Il me déclara fort absolument qu'il ne me la rendroit pas; mais en échange il me fit offre de l'Oiseau qu'il avoit ordonné de bruler et qu'il n'étoit pas impossible de remettre à flot. Je lui répondis que je me croyois en droit de réclamer la Mignonne, que cependant j'aurois moins sujet de me plaindre s'il me lassoit un des deux navires que s'il bruloit l'un et emmenoit l'autre. L'Acad. des sciences, vu la solitude de Rodrigue, avoit jugé à propos de m'associer M Thuillier Professeur de Mathématique à Paris, pour m'aider dans mes observations en qualité d'adjoint, et lui en avoit fait expédier un Acte de 29 de Novembre 1760. Le Sr. Fletcher a pris copie de cet Acte, et a laissé M Thuillier libre, non sans avoir fait beaucoup de difficultés. J'étois alors malade, et il me retrait encore bien des effet sur la Mignonne. M Thuillier fut à bord pour les reconnoitre, et les reclamer. On ne permit pas qu'il visitat ma chambre, sous prétexte qu'elle avoit occupée depuis moi par le Capitaine: et la pluspart de ces effets m'ont été retenus. Le lendemain, M. Thuillier n'étant pas encore de retour, le Sieur Fletcher est venu à terre, et a ordonné de bruler l'Oiseau. Il m'a réitéré l'offre qu'il m'avoit faite la veille, mais en m'apposant la condition que je ferois ensuite conduire ce batiment à Madras, aux dépens de la compagnie ou aux miens, et que je lui donnerois par écrit ma parole d'honneur que cette condition seroit exécutée: elle étoit trop ridicule pour être acceptée. J'offris seulement, puisque le batiment devoit être brulé, d'engager ma parole d'honneur qu'il le seroit aussitôt après mon retour à l'isle de France. Cet offre ne fut pas du gout de Sr. Fletcher: il avoit résolu de nous abandonner à la merci de la Providence. L'Oiseau n'étoit dans le port que du jour qui avoit précédé l'arrivée de la Calapate: il étoit chargé de vivres pour Rodrigue, qui n'en produit points on réprésenta au Sr. Fletcher, que ces vivres seroient même modiques pour la nombre de ceux qui resteroient dans l'isle: il donna deux heures pour les retirer. Mais nous n'avons ni canot ni pirogue pour aider à décharger l'Oiseau: nous n'avions personne à employer à ce travail: touts nos gens étoit échappés dans les bois: il nous avoit été impossible de les faire revenir: et Le Sr. Fletcher refusa de nous procurer de la part de ses gens un secours qui nous étoit absolument nécessaire. Les Officiers de l'Oiseau se mitent à l'eau, et sauverent quelques sacs de ris, le mieux qu'il leur fut possible. Au bout de deux heures on mit le feu à l'Oiseau: et le Sr. Fletcher retourna triomphant à bord de sa nouvelle prise. Il avoit décidé que ce que l'on trouveroit dans l'isle seroit partagé en deux également, qu'une partie seroit transportée sur les vaisseaux, et qu'on nous laisseroit l'autre pour notre subsistance. Soit que ces ordres aient été changés, soit qu'ils aient été mal exécutés, notre habitation a été absolument ravagée, presque toute la volaille enlevée, toutes les chevres excepté une seule, conduites à bord, les arbres absolument depouillés de leur fruits, le jardin potager livré au pillage &c. Enfin la Mignonne étant parti le 3 et la Calapate le 5 de Juillet, nous restâmes environ 80 personnes sur l'isle, reduits a la seule boisson ignoble de l'eau, n'aiant pour toute provisions que 6 à 700 livres de ris, et quelques farines qui avoient échappées à la recherche du Sr. Fletcher. Ces provisions diminuoient beaucoup, lorsqu'il nous vint un secors inespéré, d'où nous ne l'attendions pas. La Baleine Frégate Angloise, et la Drake espece de Bombardiere vinrent mouiller en rade de Rodrigue quinzaine jours après le Sieur Fletcher. Nous reconnuines la politesse et l'humanité de la nation Angloise dans les procédés de M. Phillipe Affleck Capitaine de la Baleine et de touts les autres Officiers de ces deux Navires. Les secours qu'ils nous laisserent n'étoient pas cependant intarissables; ils alloient finir, lorsque le Volant Corvette Françoise nous tira le 8 Septembre du lieu de notre exile, ainsi que les Officiers de l'Oiseau et de ceux de la Mignonne que Le Sr. Fletcher avoir relachés avant son départ. Nous étions alors, sérieusement occupés à construire une chaloupe, qui auroit été notre unique refuge pour sortir de Rodrigue, si nous n'eussions trouvé une occasion plus voisine et moins périlleuse.

Sur l'exposé de ces faits, je me flatte, My Lords, que vous voudrez bien déclarer la prise de la Mignonne injuste, et ordonner en conséquence qu'elle soit restituée. Quant à ce qui regarde la conduite illégitime et presque cruelle du Sieur Robert Fletcher a mon égard, je m'en refera à ce que votre équité vous dictera de prononcer à cet égard.

Fait en l'isle de France, le 15 7bre 1761.

Nous soussignés, jadis Officiers sur le Mignonne et sur l'Oiseau, certifions que les faits contenus dans la presente plainte sont conformes à l'exacte vérité. fait en l'isle de France.

Nous soussignés Officiers sur le Volant certifions avoir trouvé M Pingré, M. Thuillier, MM les Officiers de la Mignonne et de l'Oiseau, prets à manquer des secours les plus nécessaires, lorsque nous sommes arrivés à Rodrigue le 6 de ce mois. Fait en l'isle de France, les jours et un que dessus.

Source: Bibl. Ste. Gen., Pingré MSS, 1977, fols. 21-22.

[D]

> By the Commissioners
> for executing the office
> of Lord High Admiral of
> Great Britain & Ireland
> etc.

Whereas the Academy of Sciences at Paris have appointed two of their members to proceed to different parts of the world, to observe the expected *Transit of Venus over the Sun*; one of whom, the Bearer, *Monsieur Pingrè*, is to make such observation in *The Island*

Rodrigue in the East Indies; and whereas it is necessary that the said Monsieur Pingrè should not meet with any Interruption either in his passage to or from that Island, you are hereby most strictly required and directed *not to molest his person* or Effects upon any Account, but to suffer him to proceed without delay or Interruption in the Execution of his design. Given under our hands the 25th of November 1760.

/s/ Anson
/4 further illegible signatures/

To

The Respective Captains and
Commanders of His Majesty's
Ships & Vessels & Commanders
of Privatiers.

By Command of their Lordships.

/signature illegible/

APPENDIX IV

The Sun's Equatorial Horizontal Parallax and its Mean Distance from the Earth.

Parallax	Mean Distance (in miles)
8.0′	102,173,020
8.1	100,911,600
8.2	99,680,990
8.3	98,480,020
8.4	97,307,620
8.5	96,162,840
8.6	95,044,640
8.7	93,952,200
8.8	92,884,540
8.9	91,840,920
9.0	90,820,470
9.1	89,822,420
9.2	88,846,100
9.3	87,890,780
9.4	86,955,740
9.5	86,040,440
9.6	85,144,200
9.7	84,266,400
9.8	83,406,540

Parallax	Mean Distance (in miles)
9.9	82,564,060
10.0	81,738,420

Source: R. A. Proctor, *Transits of Venus* (New York: R. Worthington and Co., 1875), p. 236.

APPENDIX V

A Copy of Delisle's Distribution List for the *Mappemonde* and its Accompanying Memoir.

Distribuer à l'Académie

de Mairan
de Fouchi
Maraldi
Fontaine
Pingré
de Thury
Vaucanson
la Caille
la Lande
du Tillet
la Condamine
Montigni
Clairaut
Courtivron
P. Boscovich

le chev. de Lorenz
de Malesherbes
à M. de Mairan pour un ami
à M. Brache
au Comte de St. Florentin
M. Satti
M. de la Porte
M. Chabert
M. de la Post
à Mrs. Guerin et de la Tour
à M. Baer qui l'a envoié en Suède
M. Questore
M. de Montucla

En Hollande

à M. le Comte de Benkionek Seigneur de Rhoon et de Pendrecht
à la Haye

En Hollande

à M. Vosonaer à la Haye
à M. Sutoffs à Leyde
à M. Heorblert à la Haye ou à Leyde
à M. Nicolas Struick à Amsterdam

au P. Becaria à Turin
au Chevalier d'Osorio premier ministre du Royal Sardaigne à Turin
au Marquis de Fleuri à Turin
à M. Daniel Bernoulli à Basle
à Milan
au P. Ximenes à Florence

au baron de Kruikembourg envoié de l'électeur Palatine à Paris par lui j'ay envoié—
au S.A.S. l'Electeur Palatine à Mannheim
au R. P. Mayer à Heidelberg
au R. P. Huberti à Wurtzbourg
au R. P. Kraz à Ingolstad
au P. Goldtroner à Potting
au P. Liesganig à Vienne
au P. Hell à Vienne
au P. Weiss à Tyrnau en Hongrie

En Hollande

à M. Narbonnier commissionaire de vin à Dijon
au P. Christian Rieger
au P. Caspar Saguer
au Marquis de la Enferada

le 17 juillet j'ay remis 8 exemplaires de la mappemonde et du memoir en feuilles pour etre envoies à Nuremberg et distribuer de la sorte en Allemagne
à Goetting à Mrs. Mayer, Lowitz et Frantz
à Leipsic 2 exempl. l'un pour M. Heinsius et l'autre pour les Actes Leipsians
à Berlin 2 ex. l'un pour M. Euler et l'autre pour la Societé de Prusse
à Wittembourg pour M. Bose

M. L'abbé Motet Secristad de P. Benost
l'abbé Barthelemi
le baron Scheffer en Suede 2 examplaires envoier à M. Wargentin pour etre distribuer aux astronomes
un exemplaire pour Constantinople
au Suedois qui devoit partir le lendemain pour aller trouver l'ambassadeur de Suede à Constantinople

2 exempl. à Petersburg pour l'Imperatrice
4 exemplaires pour M. de Saintes et Saillant
à M. le General Betskoy à Paris
trois exemplaires vendus à M. de Mairan
3 ex. pour Turin vendus au libraire Nicolas van Daaten de la Haye
10 ex. pour l'Angleterre vendus à Mrs Saintes et Saillant

le P. Berthier de l'oratoire
l'abbé Magailhans
M. de la place pour le mercure de France
l'abbé de la porte pour l'observateur Litteraire
le Conseiller du grand Conseil M. Baudoin
l'abbé des petites affiches aubert

M. Bonamy pour le journal de verdun
M. Cutrion
M. Freron auteur de l'année Litteraire
le P. Breganaki confesseur de la Reine

M. d'Huttland
M. Allard
M. Mith ami de M. Libour
M. Saverien
l'abbé de la Dinke
M. Tennevat
le P. Berario à Lyon
le P. Pezinas à Marseille
le P. la Grange à Marseille
M. Guillembret à Montpellier
le Prince Galitzin
le Prince de Conti
à M. de Bologne
à M. le Comte d'Afry
M. Menard
Prince de Saler
pour M. Zanotti à Bologne
à Mgr. Cerati à Pise
au P. Tinsi à Pise

à M. Berger
M. Moran pour les arch.
de l'Academie
à l'abbé de la Caille
M. Chappe
M. Desparireuse
le marquis de Turbil
M. Martin
M. Monscarille
M. Jaurat
au P. Rejnau
au P. Bertiere
M. le Merre
le Cardinal de Luynes
le Monnule, medicin
P. la Tour
M. Nedham

3 exemp. pour la Soc. R. de Londres donner au P. Boscovich
Passemand
l'Etude de M. le Monnier, astronome
2 exempl. à M. Nedham pour l'Angleterre
M. Birch
M. Bager
P. Domnair
P. de Merville
à Portugal
M. Dirk Klinkenberg
25 misc.

Source: Obs. de Paris, Delisle mss, A. 3-12 (34.b.119) and A. 3-12
(34.b.120).

211

BIBLIOGRAPHY AND INDEX

MANUSCRIPT SOURCES

(Abbreviations for frequently cited MSS and
Collections appear in parentheses after the full title)

Boston

Massachusetts Historical Society
Winthrop Papers, Bowdoin-Temple, XXII, 1,2,5

London

British Museum
Le Gentil, Farquhar Funds, VI
The Royal Society
Council Minutes of the Royal Society (CMRS)
Journal Book of the Royal Society (JBRS)
Letters and Papers of the Royal Society, 1741-1806
Maskelyne Correspondence, 244 (Ma)
Miscellaneous Manuscripts, x (Roy. Soc. Misc. MSS)

Paris

Archives de l'Académie des Sciences (Arch. Acad. Sc.)

Dossier biographiques de l'Académie royale des sciences

Registres des procès-verbaux de l'Académie royale des sciences, 1740-1770 (Proc. Verb.)

Archives de l'Observatoire de Paris (Obs. de Paris)

Cassini IV; Les fastes d'astronomie, D.1,25-28.

Delisle: Astronomes, A.1, 1-10,2-11; Diamètre apparant du Soleil et de la Lune, A.2,11,25a; Mémoires sur le passage de Vénus, A.3,12,34b (4.b.6); Passages de Mercure, A.6, 3-6,55-58 and A.6, 7-8,59-60; Observations et calculs sur la parallaxe du Soleil, A.6,9,61; Astronomes, A.7,10; Correspondance, B.1, 1-8 (140 or XIV-XV) and E.1,13 (146); Voyage en Sibérie, E.1,8 (124); Inventaire des livres d'astronomie, B.5, 14.

Journal des observations faites à l'Observatoire de Paris, 1683-1798, D.3,1-30 and D.4,1-29.

La Caille: Observations faites au Collège Mazarin, au Cap, etc. 1742-1762, C.3,17-25.

Bibliothèque et Archives de la Dépôt des Cartes et Plans de la Marine (Dép. Marine)

Administration, Mémoires, Correspondance, 1751-69: 111-11; 1770-73: 111-111

215

Papiers et Manuscrits de M. Delisle, Vol. 115,XVI-6; XVI-7; 115,XVI-8.

Bibliothèque de la Chambre des Députés (Chamb. Dép.)

Bibliothèque Nationale (Bib. Nat.)

Delisle: MSS, Fr. 9678, f31; lettres, 22227

La Caille: 12275 Français

Lalande: 12273; lettres 12305, 22147, n.a.4073; papiers, 12271-12275

Le Gentil: Arch. Colon., Corresp. Madag., carton II, dossier 3, pièce 2.

Le Monnier: n.a. 6197; lettres 12305, n.a. 5151

Observations et correspondance astronomiques des XVIIe et XVIIIe siècles, MSS n.a. 6197

Papiers divers de l'Académie des sciences, 5150-5153

Bibliothèque Sainte Geneviève (Bib. Ste. Gen.)

Pingré: Calculs divers, 531; Journal à bord de la Flore, 533; Observations astronomiques et lettres de l'île Bourbon, 1085; Observation du passage de Vénus, 1175; Voyage de Paris à l'île Rodrigue, 1803; Voyage aux îles de l'Amérique, 1805; Calculs astronomiques faits dans l'île Rodrigue, 1810; Passeports, lettres, etc., 1977; papiers scientifiques, 2312; Notes et mémoires sur l'astronomie, 2321; Opuscules astronomiques et géographiques, 2342 and 2540; Correspondance, 1722-1790, 2551.

Institut de France

Delambre: Papiers, XIII-XIV (2041-2042)

La Gournerie: II (2118)

Service Hydrographique de la Marine

Journal de voyage de Pingré, 2537

PRINTED SOURCES

The lengthy titles of certain memoirs and transactions appear frequently enough in the items below to warrant abbreviation. The following short titles have therefore been adopted:

Comptes rendus for *Comptes rendus hebdomadaires des sciences de l'Académie des sciences* (Paris)
Histoire for *Histoire de l'Académie royale des sciences*
Mémoires for *Mémoires de l'Académie royale des sciences*

BIBLIOGRAPHY

Mémoires par divers savans   for *Mémoires de mathématique et de physique, présentés à l'académie royale des sciences, par divers savans, & lus dans ses assemblées.*

Novi Commentarii   for *Novi Commentarii Academiae Scientiarum Imperialis Petropolitanae*

Phil. Trans.   for *Philosophical Transactions of the Royal Society*

Phil. Trans. Abgd.   for *Philosophical Transactions of the Royal Society*, abridged by C. Hutton, G. Shaw and R. Pearson, London: 1809

Aepino, F. V. T.   "De effectu Parallaxeos in transitu Planetarum sub Sole," *Novi Comentarii*, x (1764), 433-472.

Airy, G. B. (ed.).   *Account of Observations of the Transit of Venus 1874, December 8, Made under the Authority of the British Government: and of the Reduction of the Observations.* London: 1881. Has useful material on the eighteenth-century stations.

Almeida, T. de.   "Observation du passage de Vénus sur le disque du Soleil, faite à la ville de Porto en Portugal, en l'année 1761," *Mémoires par divers savans*, vi (1774), 352.

Alzate, B.   *Suplemento a la famosa observacion del transito de Venus por el disco del sol.* Mexico: 1769.

Andrieux, L.   *Pierre Gassendi.* Paris: 1927.
A biographical study.

The Antidote;   or an *Enquiry into the Merits of a Book, Entitled A Journey into Siberia, Made in MDCCLXI, in Obedience to an order of the French King, and Published, with approbation by the Abbé Chappe d'Auteroche, of the Royal Academy of Sciences: In which many Essential Errors and Misrepresentations are pointed out and confuted; and many interesting Anecdotes added, for the better Elucidation of the several Matters necessarily discussed.* London: 1772.
Catherine II's refutation of Chappe d'Auteroche's interpretation of Russian life.

Armitage, A.   "The Astronomical Work of Nicolas-Louis de La Caille," *Annals of Science*, xii, 3 (September 1956), 163-191.

——————. "Chappe d'Auteroche: A Pathfinder for Astronomy," *Annals of Science*, x, 4 (December 1954), 277-293.

——————. "The Pilgrimage of Pingré. An Astronomer-Monk of Eighteenth-century France," *Annals of Science*, ix, 1 (March 1953), 47-63.

Armstrong, W. C.   *Lord Stirling at the Telescope.* New Brunswick, N. J.: 1907.
Concerns 1769 transit; of limited value.

217

"Astronomie," *L'Avantcoureur*, xix (26 Mai 1760).
On the Delisle-Trébuchet controversy.

Astrophilus. "History of the Transit of Venus in 1639," *Gentleman's Magazine*, xxxi (May 1761), 222-225.
About Horrox and Crabtree.

——————. "Letter to M. Urban," *Gentleman's Magazine*, xxxi (February 1761), 77-79.
On Halley's errors.

Aubert, A. "Transit of Venus over the Sun, observed June 3, 1769," *Phil. Trans.*, lix, 51 (1769), 378ff.

Audiffredi, G. B. *De Solis parallaxi*. Rome: 1766.

——————. *Investigatio parallaxis Solaris*. Rome: 1765.

——————. *Transitus Veneris ante solem observati Romae etc. Expositio historico-astronomica*. Rome: 1762.

Babinet. "Sur le diamètre apparent de la planète Vénus et sur de nouvelles présomptions contre l'exactitude de la parallaxe du Soleil déduites des derniers passages de 1761 à 1769," *Comptes rendus*, 44 (1857), 526-528.

Bailey, J. E. *The Writings of Herrox [sic] and Crabtree, Observers of the Transit of Venus, Nov. 23, 1639*. London: 1882.

Bailly, J. S. *Histoire de l'astronomie ancienne depuis son origine jusqu'a l'établissement de l'école d'Alexandrie*. Paris: 1775.

——————. *Histoire de l'astronomie moderne depuis la fondation de l'école d'Alexandrie, jusqu'à l'époque de MDCCLXXXII*. Paris: 1782.

Banks, J. *Journal of the Right Hon. Sir Joseph Banks During Captain Cook's First Voyage in H.M.S. Endeavour in 1768-71*. London: 1896.

Bayley, W. "Astronomical Observations made at the North Cape, for the Royal Society," *Phil. Trans.*, lix, 36 (1769), 262-272.

Beaglehole, J. C. (ed.). *The Journals of Captain James Cook On His Voyages of Discovery, The Voyage of the Endeavour 1768-1771*. Cambridge: 1955.

Bell, L. *The Telescope*. New York: 1922.

Bergman, T. "Observations made on the same Transit at Upsal in Sweden," *Phil. Trans.*, lii, 43 (1761), 227ff.

Bernoulli, J., III. *Lettres astronomiques où l'on donne une idée de l'état actuel de l'astronomie pratique dans plusieurs villes de l'Europe*. Berlin: 1771.

——————. *Recueil pour les astronomes*. Berlin: 1771.

Bertrand, J.-L. F. *L'Académie et les académiciens, de 1666 à 1793*. Paris: 1869.
Useful for some of the financial aspects of the Academy's history.

Bevis, J. "Observations of the last Transit of Venus, and of the Eclipse of the Sun the next Day; made at the House of Joshua Kirby, Esq. at Kew," *Phil. Trans.*, LIX, 25 (1769), 189-191.

————. "Of the Transit of Mercury over the Sun, Oct. 31, 1736, and Oct. 25, 1743," *Phil. Trans. Abgd.*, VIII, 725-726.

Biddle, O. "An account of the Transit of Venus over the Sun, June 3d 1769, as observed near Cape Henlopen, on Delaware," *Transactions of the American Philosophical Society*, I, 83-91.

———— and J. Bayley. "Observations of the Transit of Venus over the Sun, June 3, 1769; made by Mr. Owen Biddle and Mr. Joel Bayley, at Lewestown, in Pennsylvania," *Phil. Trans.*, LIX, 59 (1769), 414-421.

Bigourdan, G. *L'Astronomie à Beziers; l'observatoire, la querelle Cassini-Lalande.* Paris: 1927.

————. *Histoire de l'astronomie d'observation et des observatoires en France.* Paris: 1918.

————. "Les instruments et les observations de Bailly au Louvre. L'Observatoire de l'abbaye de Sainte-Geneviève, à Paris," *Comptes rendus*, 170 (1920), 865-871.

————. "Les instruments et les travaux de l'Observatoire de Sainte-Geneviève," *Comptes rendus*, 170 (1920), 1222-1228. Material on Pingré.

————. *Inventaire général et sommaire des manuscrits de la bibliothèque de l'Observatoire de Paris.* Paris: n.d.

Bliss, N. "An Account of a printed Memoir, in Latin, presented to the Royal Society intitled, De Veneris ac Solis congressu observatio, habita in astronomica specula Bononiensis Scientiarum Instituti, die, 5 Junii 1761. Auctore Eustachia Zanotto, ejusdem Instituti Astronomo, ac Regiae utriusque Londiensis et Berolinensis Academiac Socio," *Phil. Trans.*, LII, 63 (1761), 399ff.

————. "A Second Account of the Transit of Venus over the Sun, June 6, 1761," *Phil. Trans.*, LII, 45 (1761), 232ff.

————. "Observations on the Transit of Venus over the Sun, June 6th 1761," *Phil. Trans.*, LII, 32 (1761), 173ff.

Bode, J. E. *Deutliche Abhandlung, nebst einer allgemeinen Charte von dem bevorstehenden merkwürdigen Durchgang der Venus durch die Sonnenscheibe am 3ten Junii dieses 1769sten Jahr.* Hamburg: 1769.

Boquet, F. "Le bicentenaire de Lacaille," *L'Astronomie*, XXVII (1913), 457-473.

Boscovich, R. J. *A Theory of Natural Philosophy, with a Short Life of Boscovich.* Chicago: 1922. The brief biography is useful.

————. "On the next approaching Transit of Venus over the Sun," *Phil. Trans.*, LI (1760), 865ff.

Boscovich, R. J. and C. Marie. *Voyage astronomique et géographique dans l'état de l'Eglise, entrepris par l'ordre et sous les auspices du Pape Benoit XIV, pour mesurer deux degrés du méridien, & corriger la carte de l'état ecclésiastique.* Paris: 1770.

Bougerel, J. *Vie de Pierre Gassendi, Prevôt de l'Eglise de Digne et professeur de mathématiques au Collège Royal.* Paris: 1737. One of the earliest biographies; to be used with caution.

Bouillet père et fils, et de Manse. "Observation du passage de Vénus sur le disque du Soleil, faite à Bésiers, le 6 Juin, en présence de M. l'Evêque, Président de l'Académie de cette ville," *Mémoires par divers savans*, VI (1774), 124-132.

Bouin et Dulague. "Observation du passage de Vénus sur le disque du Soleil, faite à Rouen le 6 Juin 1761," *Mémoires par divers savans*, VI (1774), 43-44.

Bourriot. "Extract of a Letter, dated Paris, Dec. 17, 1770, to Mr. Magalhaens, from M. Bourriot; containing a short Account of the late Abbé Chappe's Observation of the Transit of Venus, in California," *Phil. Trans.*, LX, 50 (1770), 551-552.

Bradley, J. "Some Account of the Transit of Venus and Eclipse of the Sun, as observed at Lizard Point, June 3d, 1769," *Transactions of the American Philosophical Society*, I, 108-110.

Brasch, F. E. "John Winthrop (1714-1779), America's First Astronomer, and The Science of His Period," *Publications of the Astronomical Society of the Pacific*, XXVIII, 165 (August-October 1916), 153-170.

——————. "Newton's First Critical Disciple in the American Colonies—John Winthrop, [Jr.]," *Sir Isaac Newton, 1727-1927.* Baltimore: 1928.

——————. "The Newtonian Epoch in the American Colonies (1680-1783)," *Proceedings of the American Antiquarian Society*, XLIX, n.s. (1949), 314-332.

——————. "The Royal Society of London and Its Influence upon Scientific Thought in the American Colonies," *The Scientific Monthly*, XXXIII (1931), 336-355, 448-469.

Brown, L. *The Story of Maps.* Boston: 1949. Has useful material on Delisle as a cartographer.

Brunet, P. *Maupertuis.* 2 vols. Paris: 1929.

Budd, T. H. *The Transit of Venus.* London: 1875. A poor popularization!

Cajori, F. (ed.). *Sir Isaac Newton's Mathematical Principles of Natural Philosophy and his System of the World.* Berkeley: 1946.

Calder, I. M. (ed.). *Letters and Papers of Ezra Stiles, President of Yale College 1778-1795.* New Haven: 1933. Contains letters on the publication of B. West's transit of Venus pamphlet.

Canton, J. "On the Transit of Venus, June 6, 1761, made in Spital Square; the Longitude of which is 4′11″ West Latitude 51°31′ 15″ North," *Phil. Trans.*, LII, 34 (1761), 182ff.

Carcani, P. N. M. *Passagio di Venere sotto il Sole osservato in Napoli nel Real Collegio delle Scuole Pie, la mattina de' 6 Giugno 1761.* Napoli: n.d.

Carré, H. *Le Règne de Louis XV* (1715-1774). Paris: 1911.

Carrington, H. (ed.). *The Discovery of Tahiti*. Hakluyt Society Publications, Series II, XCVIII, London: 1948.

Carstrom, J. D. D. *Dissertatio de Venere in Sole visa dei 6 Junii Anni 1761.* Aboae: n.d.

Carvalho, Joaquin de (ed.). *Correspondência científica dirigida a João Jacinto de Magalhães (1769-1789). Inedita ac Rediviva Subsídos para a História da Filosofia e da Ciência em Portugal.* Coimbra: 1952.

Cassini de Thury, C.-F. "Extrait des observations du passage de Vénus sur le Soleil, faites par M. l'Abbé Chappe, en 1769," *Mémoires*, 1770, pp. 83-89.

————. "Histoire abrégé de la parallaxe du Soleil," in Chappe d'Auteroche, J., *Voyage en Californie pour l'observation du passage de Vénus sur le disque du Soleil, le 3 juin 1769; contenant les observations de ce phénomène, & la description historique de la route de l'auteur à travers le Mexique.* Paris: 1772.

————. "Observation du passage de Vénus sur le disque du Soleil. Faite à l'Observatoire royale le 3 Juin 1769," *Mémoires*, 1769, pp. 29-32.

————. "Observation du passage de Vénus sur le Soleil, faite à Vienne en Autriche," *Mémoires*, 1761, pp. 409-412.

————. "Recherche de la parallaxe de Mars et de Vénus, par les observations correspondantes faites au Cap de Bonne-Espérance & à l'Observatoire de Paris," *Mémoires*, 1760, pp. 292-303.

————. "Remarques sur la conjonction de Vénus avec le Soleil, qui doit arriver le 6 Juin de l'année prochaine 1761," *Mémoires*, 1757, pp. 326-335.

Cassini IV, J.-D. *Éloge de M. Le Gentil, membre de l'Académie royale des sciences de Paris.* Paris: 1810.

————. *Mémoires pour servir à l'histoire des sciences et à celle de l'Observatoire royal de Paris, suivis de la vie de J.-D. Cassini écrite par lui-même, et des éloges de plusieurs académiciens mort pendant la révolution.* Paris: 1810.

————. *Réflections présentées aux éditeurs des futures éditions de "l'Histoire de l'astronomie au XVIIIe siècle."* Paris: n.d.

Chabert, Joseph-Bernard, Marquis de. "Mémoire sur l'avantage de la position de quelques isles de la mer de Sud, pour l'observation de l'entrée de Vénus devant le Soleil, qui doit arriver le 6 Juin 1761," *Mémoires*, 1757, pp. 49-51.

—————. "Mémoire sur la nécessité, les avantages, les objets & les moyens d'exécution du voyage que l'Académie propose de faire entreprendre à M. Pingré dans la partie occidentale & méridionale de l'Afrique, à l'occasion du passage de Vénus devant le Soleil, qui arrivera le 6 Juin 1761," *Mémoires*, 1757, pp. 43-49.

Chambers, R. *A Biographical Dictionary of Eminent Scotsmen.* Glasgow: 1853.
Useful for material on James Ferguson.

Chapin, S. L. "The Astronomical Activities of Nicolas Claude Fabri de Peiresc," *Isis*, XLVIII, 1 (March 1957), 13-29.

"Chappe d'Auteroche, Jean." *Biographie Universelle* (Michaud), VII, 492-493.

Chappe d'Auteroche, J. "Addition au mémoire précédent, sur les remarques qui ont rapport à l'anneau lumineux, & sur le diamètre de Vénus, observé à Tobolsk le 6 Juin 1761," *Mémoires*, 1761, pp. 373-377.

—————. "Extract from a Journey to Siberia, for observing the Transit of Venus over the Sun," *Gentleman's Magazine*, XXXIII (November 1763), 547-552.

—————. "Extrait du voyage fait en Sibérie, pour l'observation de Vénus sur le disque du Soleil, faite à Tobolsk le 6 Juin 1761," *Mémoires*, 1761, pp. 337-372.

—————. *A Journey into Siberia Made by Order of the King of France.* London: 1774.

—————. *Mémoire du passage de Vénus sur le Soleil; contenant aussi quelques autres observations sur l'astronomie, et la déclinaison de la boussole faites à Tobolsk en Sibérie l'année 1761.* St. Petersburg: 1762.

—————. "The same Transit at Tobolsk in Siberia," *Phil. Trans.*, LXX, 47 (1761), 254ff.

—————. *Voyage en Californie pour l'observation du passage de Vénus sur le disque du Soleil, le 3 Juin 1769; Contenant les observations de ce phénomène, & la description historique de la route de l'auteur à travers le Mexique.* Paris: 1772.

—————. *Voyage en Sibérie, fait par ordre du Roi en 1761; contenant les moeurs, les usages des Russes, et l'état actuel de cette puissance; la description géographique & le nivellement de la route de Paris à Tobolsk.* Paris: 1778.

—————. *A Voyage to California to observe the Transit of Venus.* London: 1778.

Chapple, W. "Observations of the Transit at Exeter," *Gentleman's Magazine*, XXXI (June 1761), 248 and (August 1761), 357-359.

Chaulnes, le Duc de. "Observation du passage de Vénus sur le Soleil, du 3 Juin 1769, faite à l'Observatoire avec lunette de Dollond de 3 pieds & demi," *Mémoires*, 1769, pp. 529-530.

Cipolla, L. "Astronomical Observations by the Missionaries at Pekin," *Phil. Trans.*, LXIV, 2 (1774), 31-45.

Clerke, A. M. "Charles Mason," *Dictionary of National Biography*, XII, 1302.

—————. "Nevil Maskelyne," *Dictionary of National Biography*, XII, 1299-1301.

—————. *A Popular History of Astronomy During the Nineteenth Century*. London: 1902.
Has some useful material on eighteenth-century astronomy.

Cohen, I. B. "Benjamin Franklin and the Transit of Mercury in 1753," *Proceedings of the American Philosophical Society*, 94, 3 (June 1950), 222-232.

—————. *Some Early Tools of American Science*. Cambridge: 1950.
Contains material on Winthrop.

Cohen, M. R. and I. E. Drabkin. *A Source Book in Greek Science*. New York: 1948.

Collections of the New York Historical Society (1920), Colden Papers, IV, 367-368.

Collectio omnium observationem quae occasione transitus Veneris per Solem a. MDCCLXIX per Imperium Russicum institutae fuerunt. Petropoli: 1770.

Condorcet, M. J. A. N. C. "Éloge de Cassini," *Oeuvres*. Paris: 1847-1849, III, 168-180.

—————. "Éloge de M. de Courtanvaux," *Oeuvres*. II, 456-466.

—————. "Éloge de M. Euler," *Oeuvres*. III, 1-42.

—————. "Éloge de M. Fouchy," *Oeuvres*. III, 310-327.

—————. "Éloge de M. le Cardinal de Luynes," *Oeuvres*. III, 306-310.

—————. "Éloge de M. Wargentin," *Oeuvres*. III, 120-125.

—————. "Klingenstierna," *Oeuvres*. II, 127-129.

Connoissance des temps pour l'année 1761. Paris: 1759.

Cook, J. *Captain Cook's Journal During His First Voyage Round the World Made in H.M. Bark "Endeavour" 1768-71*. London: 1893.

Cope, T. D. "Charles Mason and Jeremiah Dixon," *The Scientific Monthly*, LXII (June 1946), 541-554.

—————. "A Clock Sent Thither by the Royal Society," *Proceedings of the American Philosophical Society*, 94, 3 (June 1950), 260-268.

—————. "Collecting Source Material about Charles Mason and

Jeremiah Dixon," *Proceedings of the American Philosophical Society*, 92, 2 (May 1948), 111-114.

—————. "The First Scientific Expedition of Charles Mason and Jeremiah Dixon," *Pennsylvania History*, XII, 1 (January 1945), 3-12.

—————. "A Frame of Reference for Mason and Dixon," *Proceedings of the Pennsylvania Academy of Science*, XIX (1945), 79-86.

—————. "Mason and Dixon and Franklin," *Proceedings of the Pennsylvania Academy of Science*, XXV (1951), 167-170.

—————. "Mason and Dixon—English Men of Science," *Delaware Notes*, 22nd Series (1949), 13-32.

—————. "More about Mason and Dixon," *Proceedings of the Pennsylvania Academy of Science*, XXIII (1949), 216-217.

—————. "Some Contacts of Benjamin Franklin with Mason and Dixon and their Work," *Proceedings of the American Philosophical Society*, 95 (June 1951), 232-238.

—————. "The Stargazers' Stone," *Pennsylvania History*, VI, 4 (October 1939), 205-220.

—————. "Westward Five Degrees in Longitude," *Proceedings of the Pennsylvania Academy of Science*, XXII (1948), 145-152.

Cope, T. D. and H. W. Robinson. "The Astronomical Manuscripts which Charles Mason Gave to Provost the Reverend John Ewing During October 1786," *Proceedings of the American Philosophical Society*, 96, 4 (August 1952), 417-423.

—————. "Charles Mason, Jeremiah Dixon and the Royal Society," *Notes and Records of the Royal Society of London*, IX (October 1951), 55-78.

Cutler, W. P. (ed.). *Life, Journals and Correspondence of Rev. Manasseh Cutler, LL.D.* Cincinnati: 1888.

Daumas, M. *Les Instruments scientifiques aux XVIIe et XVIIIe siècles*. Paris: 1953.

Daval, P. "Of the Sun's Distance from the Earth deduced from Mr. Short's Observations relating to the Horizontal Parallax of the Sun," *Phil. Trans.*, LIII, 1 (1763) 1ff.

De Beer, G. R. "The Relations Between Fellows of the Royal Society and French Men of Science when France and Britain were at War," *Notes and Records of the Royal Society*, IX, 2 (May 1952), 244-299.

Degloss, L. and J. Lang and H. Stoker. "Observations made at Dinapoor, June 4, 1769 on the Planet Venus, when passing over the Sun's Disk, June 4, 1769, with three different Quadrants, and a Two-Foot reflecting Telescope," *Phil. Trans.*, LX, 23 (1770), 239ff.

Delambre, J. B. "Alexandre-Gui Pingré," *Biographie Universelle* (Michaud) XXXIII, 364-366.

Delambre, J. B. "Discours prononcé aux obsèques de Joseph-Jérôme de Lalande le lundi 6 Avril 1807," *Le Moniteur*, 102 (Avril 1807), 12ff.

——————. *Histoire de l'astronomie au XVIIIe siècle*. Paris: 1827.

Delaunay, C.-E. "Note sur la parallaxe du Soleil," *Comptes rendus*, 45 (1867), 839-841.

——————. "Notice sur la distance du Soleil," *Recueil de Mémoires, rapports et documents relatifs à l'observation du passage de Vénus sur le Soleil*. Paris: 1874.
Primarily concerned with nineteenth-century transits, this article nevertheless has much on the eighteenth-century transits.

——————. "Nouvelle note sur la parallaxe du Soleil," *Comptes rendus*, 45 (1867), 976-979.

——————. "Sur la parallaxe du Soleil," *Comptes rendus*, 45 (1867), 876-877.

"Delisle, Joseph-Nicolas," *Biographie Universelle* (Michaud), x, 335.

Delisle, J.-N. "Extrait d'une lettre de M. Delisle, écrite de Petersbourg le 24 Août 1743, & adressée à M. Cassini, servant de supplément au Mémoire de M. Delisle, inséré dans le volume de 1723, p. 105, pour trouver la parallaxe du Soleil par le passage de Mercure dans le disque de cet astre," *Mémoires*, 1743, pp. 419-428.

——————. "A Letter from M. de l'Isle, of the Royal Academy of Sciences at Paris, to the Rev. James Bradley, D.D., Dated Paris, Nov. 30, 1752," *Phil. Trans.*, XLVIII (1754), 512ff.

——————. "Lettre de M. de Lisle, Professeur royal . . . de l'Académie royale des sciences à M. . . . sur les tables astronomiques de M. Halley," *Mémoires de Trévoux* (Février 1750), pp. 377-380.

——————. *Lettre de M. De L'Isle, . . . sur les tables astronomiques de M. Halley*. Paris: 1749.

——————. "Mémoire sur le diamètre apparant de Mercure, et sur le temps qu'il emploie à entrer & à sortir du disque du Soleil dans les conjonctions écliptiques," *Mémoires*, 1753 (Amsterdam: 1762), pp. 366-375.

——————. *Mémoires pour servir à l'histoire et aux progrès de l'astronomie, de la géographie et de la physique, recueillie de plusieurs dissertations lues dans les assemblées de l'Académie royale des sciences de Paris et celle de Saint Petersbourg, qui n'ont point encore été imprimées, comme aussi de plusieurs nouvelles, observations et réflexions rassemblées pendant plus de 25 années*. Saint Petersbourg: 1738.

——————. "Observations des diamètres apparens du Soleil, faites à Paris les années 1718 & 1719, avec des lunettes de différentes longueurs; et réflexions sur l'effet de ces lunettes," *Mémoires,* 1755 (Amsterdam: 1776), pp. 215-254.

—————. "Observations du passage de Mercure sur le disque du Soleil, le 6 Novembre 1756; avec des réflexions qui peuvent servir à perfectionner les calculs de ces passages, & les élémens de la théorie de Mercure déduites des observations," *Mémoires,* 1758, pp. 134-154.

—————. "Observation du passage de Mercure sur le Soleil; faite à Paris dans l'Observatoire royal, le 9 Novembre 1723, au soir," *Mémoires,* 1723, pp. 306-343.

—————. "Sur le dernier passage attendu de Mercure dans le Soleil, et sur celui du mois de Novembre de la présente année 1723," *Mémoires,* 1723, pp. 105-110.

Denny, M. "The Royal Society and American Scholars," *Scientific Monthly,* 65 (1947), 415-427.

Devic, J.-F.-S. *Histoire de la vie et des travaux scientifiques et littéraires de J.-D. Cassini IV, ancien directeur de l'Observatoire.* Clermont (Oise): 1851.

Dexter, F. B. *Extracts from the Itineraries and other Miscellanies of Ezra Stiles, D.D., LL.D. 1755-1794 with a selection from his Correspondence.* New Haven: 1916.
Stiles' observations of the transit.

—————. (ed.). *The Literary Diary of Ezra Stiles, D.D., LL.D., President of Yale College.* New York: 1901.
Refers to the transit of Venus observations.

Dixon, J. "Observations made on the Island of Hammerfost, for the Royal Society," *Phil. Trans.,* LIX, 35 (1769), 253-262.

Doublet, E. "Le Bicentenaire de l'Abbé de la Caille," *Actes de l'Académie nationale des sciences, belles-lettres et arts de Bordeaux,* 4e série, 2 (1914-1915), 179-227.

—————. "Correspondance échangée de 1720 à 1739 entre l'astronome J-N Delisle et M. de Navarre," *Actes de l'Académie nationale des sciences, belles-lettres et arts de Bordeaux,* 3e Série, 92 (1910), 5-87.

Doz, Don V. "A Short Account of the Observations of the late Transit of Venus, made in California, by Order of his Catholic Majesty," *Phil. Trans.,* LX, 49 (1770), 549-550.

Dreyer, J. L. E. *A History of Astronomy from Thales to Kepler.* 2nd edn.; New York: 1953.

Dubois, E. P. *Les passages de Vénus sur le disque Solaire, considérés au point de vue de la détermination de la distance du Soleil à la Terre.* Paris: 1873.
Contains historical material on transits of 1761 and 1769. Excellent technical exposition.

Dubois, E. P.   "Sur l'influence de la réfraction atmosphérique relative à l'instant d'un contact dans un passage de Vénus," *Comptes rendus*, 76 (1873), 1526-1530.

Dunkin, E.   *Obituary Notices of Astronomers*. London: 1879.

Dunn, S.   "A Determination of the Exact Moments of Time when the Planet Venus was at External and Internal Contact with the Sun's Limb, in the Transits of June 6th, 1761, and June 3d, 1769," *Phil. Trans.*, LX, 9 (1770), 65-73.

—————.   "Observations of the Planet Venus, on the Sun's Disk, June 6, 1761; and Certain Reasons for an Atmosphere about Venus," *Phil. Trans.*, LII, 35 (1761), 184ff.

Duséjour, D.   "Calculs des passages de Vénus sur le disque du soleil, des 6 Juin 1761, et 3 Juin 1769," *Traité analytique des mouvemens apparens des corps célestes*. Paris: 1786, 451-491.

Dymond, J. and W. Wales.   "Observations on the State of the Air, Winds, Weather, &c. made at Prince of Wales' Fort, on the North-West Coast of Hudson's Bay, in the Years 1768 and 1769," *Phil. Trans.*, LX, 14 (1770), 137-178.

Early Proceedings of the American Philosophical Society for the Promotion of Useful Knowledge Compiled by one of the Secretaries from the Manuscript Minutes of Its Meetings from 1744 to 1838. Philadelphia: 1884.

Ebeling, J. P.   (trans.).   *Le Gentils Reisen in den indischen Meeren in den Jahren 1761 bis 1769 und Chappe d'Auteroche Reise nach Mexico und Californien im Jahre 1769 aus dem Französischen, nebst Karl Millers Nachricht von Sumatra und Franziscus Masons Beschreibung der Insel St. Miguel aus dem Englischen*. Hamburg: 1781.

Eichorn, J. A.   *Beschreibung des Durchgangs der Venus durch die Sonnenscheibe, als eine sehr seltene und weil die Welt stehet nur einmal bemerkte Himmelsbegebenheit*. Nürnberg: 1761.

—————.   *Die Erscheinung der Venus in der Sonnenscheibe, als eine sehr seltene und weil die Welt stehet nur einmal bemerkte Himmelsbegebenheit*. Nürnberg: 1761.

"Éloge de M. de l'Isle," *Histoire*, 1768, pp. 167-183.

"Éloge de M. de l'Isle," *Nécrologe des hommes célèbres de France par un société de gens de lettres*. Paris: 1770, V, 1-86.

"Éloge de M. l'Abbé Chappe," *Histoire*, 1769, pp. 163-172.

"Éloge de M. l'Abbé Chappe," *Nécrologe des hommes célèbres de France par un société de gens de lettres*. Paris: 1771, VI, 133-157.

"Éloge de M. l'Abbé de la Caille," *Histoire*, 1762, pp. 354-383.

"Éloge de M. le Cardinal de Luynes," *Histoire*, 1788, pp. 33-36.

"Éloge de M. le duc de Praslin," *Histoire*, 1785, pp. 137-155.

"Éloge de M. Wargentin," *Histoire*, 1783, pp. 128-132.

"Elogy of Dr. Halley," *Gentleman's Magazine*, XVII (October 1747), 455-458 and (November 1747), 503-547.

Elton, L. *Imperial Commonwealth*. New York: 1946.

Encke, J. F. *Die Entfernung der Sonne von der Erde aus dem Venus- durchgang von 1761 hergeleitet*. Gotha: 1822.
Though far from complete, one of the best secondary sources on the transit of 1761.

—————. "Ueber den Venusdurchgang von 1769," *Abhandlun- gen der Koeniglich-preussischen Akademie der Wissenschaften, Mathematische Klasse*, 1835, pp. 295-309.

—————. *Der Venusdurchgang von 1769 als Fortsetzung der Ab- handlung über die Entfernung der Sonne von der Erde*. Gotha: 1824.
A major source on the transit of 1769.

Eneström, G. "P. W. Wargentin und die sogenannte Halley'sche Methode," *Abhandlungen zur Geschichte der Mathematischen Wissenschaften*, IX (1899), 81-95.

Enriques, F. and G. de Santillana. *Mathématiques et astronomie de la période Hellénique*. Paris: 1939.

Euler, L. "A Deduction of the Quantity of the Sun's Parallax from the Comparison of the several Observations of the late Transit of Venus, made in Europe, with those made in George Island in the South Seas," *Phil. Trans.*, LXII, 9 (1772), 69-76.

—————. "Expositio methodorum cum pro determinanda paral- laxi Solis, ex observato transitu Veneris per Solem, tum pro inueniendis longitudinibus locorum super terra ex observationi- bus eclipsium Solis, una cum calculis et conclusionibus inde de- ductis," *Novi Commentarii*, XIV, 2 (1769), 321-554.

Evans, D. S. "La Caille: 10,000 Stars in Two Years," *Discovery*, XII, 10 (October 1951), 315-319.
On La Caille's work at the Cape of Good Hope.

Ewing, J. "An Account of the Transit of Venus over the Sun, June 3d, 1769, and of the Transit of Mercury, Nov. 9th, both as ob- served in the State-House Square, Philadelphia," *Transactions of the American Philosophical Society*, I, 39-83.

—————. "Calculation of the same for the City of Philadelphia," *Transactions of the American Philosophical Society*, I, 5.

Fath, E. A. *The Elements of Astronomy*. New York: 1944.
A brief but useful technical exposition.

Faye, H. A. E. A. "Examen critique des idées et des observations du P. Hell sur le passage de Vénus de 1769," *Comptes rendus*, 68 (1869), 282-290.

—————. "Réponses à quelques critiques relatives à la note du 21 février sur la parallaxe du soleil," *Comptes rendus*, 92 (1881), 1071-1074.

Faye, H. A. E. A. "Sur la mesure de la distance du Soleil à la Terre," *Comptes rendus*, 53 (1861), 525-529.

――――. "Sur la parallaxe du Soleil," *Comptes rendus*, 92 (1881), 375-378.

――――. "Sur les passages de Vénus et la parallaxe du Soleil," *Comptes rendus*, 68 (1869), 42-49.

Ferguson, A. (ed.). *Natural Philosophy through the Eighteenth Century and Allied Topics.* London: 1948.
Essays by various hands on different branches of eighteenth-century science.

Ferguson, J. *Astronomy Explained Upon Sir Isaac Newton's Principles. . . . To which are added A Plain Method of Finding the Distances of all the Planets from the Sun, by the Transit of Venus over the Sun's Disc, in the Year 1761. An Account of Mr. Horrox's Observation of the Transit of Venus in the Year 1639: and of the Distances of all the Planets . . . as deduced from the Transit in the Year 1761.* 5th edn.; London: 1772.

――――. "A delineation of the Transit of Venus expected in the Year 1769," *Phil. Trans.*, LIII, 8 (1763), 30.

――――. *A Plain Method of Determining the Parallax of Venus, by her Transit over the Sun: and from thence by Analogy, the Parallax and Distance of the Sun, and of all the Rest of the Planets.* London: 1761.

Ferner, B. "An Account of the Observations on the same Transit made in and near Paris," *Phil. Trans.*, LII, 41 (1761), 221.

――――. "Extract of a Letter to the Reverend Nevil Maskelyne, Astronomer Royal," *Phil. Trans.*, LIX, 57 (1769), 404-406.

――――. *Resa I Europa* (1758-1762). Uppsala: 1956.

Ferrer, J. J. de. "On the Determination of the Parallax of the Sun from the Observations of the Transit of Venus over his Disk, June 3, 1769," *Memoirs of the Royal Astronomical Society*, V (1832), 253-296.
An excellent summary.

Fisher, T. "Letter on the transit of Venus," *Gentleman's Magazine*, XXVIII (August 1758), 367-368; XXIX (January 1759), 23-26.

Flaugergues, H. "Observations inédites faites à l'Observatoire de Montpellier par MM. de Ratte, Tandon, Romieu, Poitevin, Dubousquet, etc., membres de la Société royale des sciences de la même ville," in Zach, *Correspondance astronomique, géographique, hydrographique et statistique.* Geneva: 1818, Vol. 1, pp. 246-248.

Fleming, D. *Science and Technology in Providence, 1760-1914.* Providence: 1952.
Contains material on the American observations of the transits.

Fleurieu, Comte de. *Voyage fait par ordre du Roi en 1768 et 1769, à différentes parties du monde, pour éprouver en mer les horloges marines inventées par M. Berthoud.* 2 vols.; Paris: 1773.

Fontenelle, B. "The Elogy of Dr. Halley," *Gentleman's Magazine,* XVII (October 1747), 455-458; (November 1747), 503-507.

Forbes, G. *The Transit of Venus.* London: 1874.
A popular study of little use.

Fortia d'Urban. *Histoire d'Aristarque de Samos.* Paris: 1823.

Fouchy, J. P. G. de. "Observation du passage de Vénus sur le Soleil, faite à la Muette au cabinet de physique du Roi, le 6 Juin 1761," *Mémoires,* 1761, pp. 96-105.

————, de Bory and Bailly. "Observation du passage de Vénus sur le Soleil, le 3 Juin 1769; & de l'éclipse du Soleil, du 4 Juin de la même année, faite au Cabinet de Physique du Roi, à Passy," *Mémoires,* 1769, pp. 531-538.

Foucoult, L. "Détermination expérimentale de la vitesse de la lumière; parallaxe du Soleil," *Comptes rendus,* 55 (1862), 501-503.
The solar parallax determined from studies of the speed of light.

Gassendi, P. *Institutio astronomica juxta hypotheses tam veterum quam Copernici & Tychonis.* Amsterdam: 1680.

————. "Mercuris in Sole visus, & Venus invisa Parisiis anno 1631 pro Voto & admonitione Keppleri: cum observatus quibusdam alliis," *Opera Omnia.* Lyon: 1658, IV, 537ff.

Gersten, C. L. "Transit of Mercury over the Sun, Nov. 5, 1743, seen at the Observatory at Giesen," *Phil. Trans. Abgd.,* IX, 307-308.

Gill, D. *A History and Description of the Royal Observatory, Cape of Good Hope.* London: 1913.
Discusses La Caille's work at the Cape.

Gill, H. V. *Roger Boscovich, S. J. (1711-1787).* Dublin: 1941.
Brief but convenient summary of his life and labors.

Gingerich, O. "Messier and his Catalogue," *Sky and Telescope,* XII, 10 (August 1953), 255-258, 265.

Graham, G. "Observations of the Transit of Mercury over the Sun, Oct. 31, 1736," *Phil. Trans. Abgd.,* VIII, 148-149.

Grandidier, G. *Bibliographie de Madagascar.* Paris: 1906-1935.
A useful guide to source material on Le Gentil.

Grant, R. *History of Physical Astronomy from the Earliest Ages to the Middle of the Nineteenth Century.* London: 1852.
A valuable source in spite of its age.

Green, C. and J. Cook. "Observations made, by appointment of the Royal Society, at King George's Island in the South Sea," *Phil. Trans.,* LXI, 43 (1771), 397-421.

Gregory, J. "Optica promota, seu abdita radiorum reflexorum & refractorum mysteria, geometricè enucleata; cui subnectitur appendix, subtilissimorum astronomiae problematon resolutionem exhibens. London: 1663," in Masères, F. *Scriptores optici; or A Collection of Tracts Relating to Optics.* London: 1823, 1-104.

Halley, E. "A New Method of Determining the Parallax of the Sun, or his Distance from the Earth," *Phil. Trans.* Abgd., VI, 1713-1723, 244-246.

————. "On the Visible Conjunctions of the Inferior Planets with the Sun," *Phil. Trans.* Abgd., III, 454ff.

Harding, L. A. *A Brief History of the Art of Navigation.* New York: 1952.

A useful outline. Points out that though Bowditch supported the method of lunar distances, it was not normally accepted.

Harkness, W. "On the Magnitude of the Solar System," *Smithsonian Institute Annual Report,* 1894, pp. 93-111.

————. "On the Relative Accuracy of Different Methods of Determining the Solar Parallax," *The American Journal of Science,* 3rd Series, 22 (1881), 375-394.

————. "On the Transits of Venus," *Nature,* 27 (1882), 114-117.

————. "The Solar Parallax and its Related Constants Including the Figure and Density of the Earth," *Washington Observations* (1885), Appendix III.

Harris, D. "Observations of the Transit of Venus over the Sun, made at the Round Tower in Windsor Castle, June 3, 1769," *Phil. Trans.,* LIX, 60 (1769), 422-431.

Hawkesworth, J. *An Account of the Voyages Undertaken by the Order of His Present Majesty for Making Discoveries in the Southern Hemisphere.* London: 1773.

Haydon, R. "Account of the same Transit," *Phil. Trans.,* LII, 37 (1761), 202.

Hayes, C. J. H., M. W. Baldwin and C. W. Cole. *History of Europe.* New York: 1949.

Heath, T. L. *Aristarchus of Samos.* Oxford: 1913.

————. *The Works of Archimedes.* Cambridge: 1897.

Heindel, R. H. "An Early Episode in the Career of Mason and Dixon," *Pennsylvania History,* VI, 1 (January 1939), 20-24.

Heinsio, G. "De Effecto parallaxis in Transitu Veneris per Solem," *Novi Commentarii,* x (1764), 501-543.

————. "De Venera in Sole visa anno 1761. d. 6 Iuni St. nov.," *Novi Commentarii,* x (1764), 473-500.

Hell, M. *Observatio transitus Veneris ante discum Solis, die 3 Junii anno 1769, Wardoëhusii, auspiciis potentissimi ac clementissimi regis Daniae et Norvegiae Christiani VII.* Hafniae: 1770.

Henderson, E. *Life of James Ferguson, F. R. S., in a Brief Auto-*

biographical Account, and Further Extended Memoir. 2nd edn.; Edinburgh: 1870.

Hewson, J. B. A History of the Practice of Navigation. Glasgow: 1951.

Hindle, B. The Pursuit of Science in Revolutionary America, 1735-1789. Chapel Hill, N. C.: 1956.

Hirst, W. "Account of Several Phaenomena observed during Ingress of Venus into the Solar Disc," Phil. Trans., LIX, 31 (1769), 228-235.

—————. "An observation of the same Transit of Venus over the Sun, June 6, 1761, at Madras," Phil. Trans., LII, 62 (1761), 396.

"Historical Chronical, November 1760," Gentleman's Magazine, xxx (November 1760), 528.
Announcement of the transit expeditions.

Holland, S. "Astronomical Observations made by Samuel Holland, Esq. Surveyor General of Lands for the Northern District of North America, and others of his Party," Phil. Trans., LIX, 33 (1769), 247-252.

Hollis, H. P. "Jeremiah Dixon and his Brother," Journal of the British Astronomical Association, 44 (June 1934), 294-299.

Holmes, M. "Captain James Cook, R.N., F.R.S.," Endeavour, VIII, 9 (January 1949), 11-17.

Hornberger, T. Scientific Thought in the American Colleges, 1638-1800. Austin, Texas: 1945.

Hornsby, T. "An Account of Observations of the Transit of Venus and of the Eclipse of the Sun, made at Shirburn Castle and at Oxford," Phil. Trans., LIX, 23 (1769), 172-182.

—————. "On the Parallax of the Sun," Phil. Trans., LIII, 55 (1763), 467-475.

—————. "On the Transit of Venus in 1769," Phil. Trans., LV, 34 (1765), 326ff.

—————. "The Quantity of the Sun's Parallax, as deduced from the Observations of the Transit of Venus, on June 3, 1769," Phil. Trans., LXI, 53 (1771), 574-579.

Horrebow, C. Dissertatio de semita, quam in Sole descripsit Venus per eundem transeunde die 6 Junii Ao. 1761. Havniae: 1761.

Horrox, J. Opera Posthuma. London: 1673.

Horsfall, J. "Observations of the late Transit of Venus," Phil. Trans., LIX, 22 (1769), 170-171.

Horsley, S. "A Computation of the Distance of the Sun from the Earth," Phil. Trans., LVII, 8 (1767), 179.

—————. "Venus observed upon the Sun at Oxford, June 3, 1769," Phil. Trans., LIX, 24 (1769), 183-188.

Housman, A. E. M. Manilii Astronomicon. London: 1903.

Houzeau, J. C. *Catalogue des ouvrages d'astronomie et de météorologie qui se trouvent dans les principales bibliothèques de la Belgique, préparé et mis en ordre à l'Observatoire royal de Bruxelles.* Bruxelles: 1878.

——————. *Vade-mecum de l'astronomie.* Bruxelles: 1882.

—————— and A. Lancaster. *Bibliographie générale de l'astronomie.* Bruxelles: 1887.

The most useful bibliography of its kind.

Humbert, P. *Un Amateur: Peiresc (1580-1637).* Paris: 1933.

——————. *L'Oeuvre astronomique de Gassendi.* Paris: 1936.

An excellent account.

Hutton, C. "Lettre du Dr. Hutton à M. le Marquis de Laplace," *Journal de physique, de chimie, d'histoire naturelle et des arts,* xc (Avril 1820), 307-312.

——————. *A Philosophical and Mathematical Dictionary.* London: 1815.

Isnard, A. "Joseph-Nicolas Delisle, sa biographie et sa collection de cartes géographiques à la Bibliothèque nationale," *Comité des travaux historiques et scientifiques. Bulletin de la section de géographie,* xxx (1915), 34-164.

Isnard, E. *Essai historique sur le chapitre de Digne et sur Pierre Gassendi, 1177-1790.* Digne: 1915.

"James Ferguson, the Astronomer," *Blackwood's Edinburgh Magazine,* 134, 314 (August 1883), 244-263.

Jardine, Lt. "Observations of the Transit of Venus and other Astronomical Observations, made at Gibraltar," *Phil. Trans.,* LIX, 45 (1769), 347-350.

Jeaurat. "Observations de l'opposition de Jupiter, du 8 Mai; du passage de Vénus au-devant du Soleil, du 3 Juin; & de l'éclipse de Soleil du 4 Juin 1769," *Mémoires,* 1769, pp. 147-152.

——————. "Observation du passage de Vénus sur le Soleil du 6 Juin 1761; & détermination de sa conjonction & de la position de son noeud," *Mémoires,* 1762, p. 570.

Jevons, W. S. *The Principles of Science: A Treatise on Logic and Scientific Method.* 3rd edn.; London: 1879.

Jones, H. S. "The Distance of the Sun," *Endeavour,* I, 1 (January 1942), 9-17.

Keir, D. L. *The Constitutional History of Modern Britain, 1485-1937.* London: 1948.

Kelly, J. *The Life of John Dolland, F.R.S., Inventor of the Achromatic Telescope.* London: 1808.

Contains many of Dolland's letters.

Klinkenberg, D. *Verhandeling, Beneffens de Naauwkeurige Algemeene en Byzondere Afbeeldingen van den Overgang der Planet*

Venus voorby de Zon, op den 6 Juny 1761 des Morgens. The Hague: 1760.

—————. *Verhandeling Over het vinden van de Parallaxis der Zon; Zynde eene Beschryving hoe de Afstand, tusschen de Zon en de Aarde kan gevonden worden door den schynbaaren weg der Planeeten Venus en Mercurius over de Zon: Nevens de Afbeeldingen van drie zulke Verschynsels, welke voorvallen zullen, het eene en Jaar 1743, de ander 1753, en de derde in't Jaar 1761.* Haarlem: 1743.

Kohler, C. *Catalogue des manuscrits de la Bibliothèque Sainte-Geneviève.* Paris: 1896.

Kordenbusch, G. F. *Die Bestimmung der denkwuerdigen Durchgänge der Venus 1761 und 1769.* Nuremberg: 1769.

Krafft, W. L. *Auszug aus den Beobachtungen welche zu Orenburg bey Gelegenheit des Durchgangs der Venus vorbey der Sonnenscheibe angestellt worden sind.* St. Petersburg: 1769.

—————. "Elementa astronomica theoriae Veneris deducta ex observatione transitus Veneris sub Sole ad sinum Hudsonis, Californiae et insula Regis Georgii instituta," *Novi Commentarii*, XVI (1771), 649-692.

Kuiper, G. P. (ed.). *The Sun.* Chicago: 1953.

La Caille, N. de. "Extract of a Letter from the Abbé de la Caille, of Paris, and F. R. S. to William Watson, M. D., F. R. S. recommending to the Rev. Nevil Maskelyne, F. R. S., discovering the Parallax of the Moon," *Phil. Trans.*, LII (1761), 21.

—————. "Mémoire sur la parallaxe du Soleil, qui résulte de la comparaison des observations simultanées de Mars & de Vénus, faites en l'année 1751 en Europe & au Cap de Bonne-Espérance," *Mémoires*, 1760, pp. 73-97.

—————. "Observation du passage de Vénus sur le disque du Soleil," *Mémoires*, 1761, pp. 78-81.

—————. *Reise nach dem Vorgebürge der guten Hoffnung.* Altenburg: 1778.

—————. "Relation abrégé du voyage fait par ordre du Roi, au Cap de Bonne-Espérance," *Mémoires*, 1751, pp. 519-536.

Lacroix, A. *Inauguration du monument de Jérôme de Lalande à Bourg-en-Bresse, le dimanche 18 Avril 1909.* Paris: 1909.

—————. *Notice historique sur les membres et correspondants de l'Académie des sciences ayant travaillé dans les colonies Françaises des Mascareignes et de Madagascar au XVIIIe siècle et au début du XIXe.* Paris: 1934.
Includes useful material on Le Gentil and Pingré.

"Lalande, Joseph-Jérome Lefrançais de," *Biographie Universelle* (Michaud) XXII, 603-612.

Lalande, J.-J. L. de. *L'Astronomie.* 2nd edn.; Paris: 1771.

—————. *Bibliographie astronomique avec l'histoire de l'astronomie depuis 1781 jusqu'à 1802.* Paris: 1803. Extremely useful.

—————. "Comparaison du passage de Vénus, observé à Bordeaux, avec les observations faites à Paris," *Mémoires*, 1769, pp. 509-512.

—————. "Détermination de la longitude et de la latitude de Vénus en conjonction, par la durée du passage observée à Stockholm," *Mémoires*, 1761, pp. 334-336.

—————. "Examen de la latitude & de la longitude de Foulpointe dans l'île de Madagascar, par les observations de M. le Gentil, discutées & calculées sur les meilleures tables," *Mémoires*, 1767, pp. 127-129.

—————. "Examen de la parallaxe du Soleil, par les observations de la plus proche distance des bords de Vénus et du Soleil, à Paris et à Rodrigue," *Mémoires*, 1761, pp. 90-92.

—————. "Examen de la plus courte distance des centres de Vénus et du Soleil, le 3 Juin 1769," *Mémoires*, 1769, pp. 543-546.

—————. "Explication du prolongement obscur du disque de Vénus, qu'on aperçoit dans ses passages sur le Soleil," *Mémoires*, 1770, pp. 406-412.

—————. "Extract of a Letter from M. de la Lande, at Paris, to the Rev. Nevil Maskelyne, F.R.S.," *Phil. Trans.*, LII, 99 (1762), 607.

—————. *Figure du passage de Vénus sur le disque du soleil, qui s'observera le 3 Juin 1769.* Paris: 1764.

—————. "Mémoire sur la parallaxe du Soleil, déduite des observations faites dans la mer du Sud, dans le royaume d'Astracan, & à la Chine," *Mémoires*, 1771, pp. 776-799.

—————. "Mémoire sur la parallaxe du Soleil, qui résulte du passage de Vénus, observée en 1769," *Mémoires*, 1770, pp. 9-14.

—————. "Mémoire sur le diamètre du Soleil qu'il faut employer dans le calcul des passages de Vénus," *Mémoires*, 1770, pp. 403-405.

—————. *Mémoire sur le passage de Vénus, observé le 3 Juin 1769; pour servir de suite à l'explication de la carte publiée en 1764.* Paris: 1772.

—————. "Mémoire sur les observations du passage de Vénus faites à Brest," *Mémoires*, 1769, pp. 546-548.

—————. "Mémoire sur les passages de Vénus devant le disque du Soleil, en 1761 et 1769, dans lequel on exprime, d'une manière générale, l'effet de la parallaxe dans les différens lieux de la Terre, pour l'entrée & pour la sortie de Vénus, soit par le calcul, soit par les opérations graphiques; avec les remarques sur l'avantage qu'il y auroit à observer la sortie, en 1761, vers l'extrémité de l'Afrique," *Mémoires*, 1757, pp. 232-250.

Lalande, J.-J. L. de. "Méthode pour trouver avec la plus grand précision le mouvement horaire de Vénus ou de Mercure dans leur passages sur le Soleil," *Mémoires*, 1762, p. 96.

—————. "Observations de M. l'Abbé Chappe, faites en Californie pour le passage de Vénus, avec les conséquences qui en résultent," *Mémoires*, 1770, pp. 416-422.

—————. "Observation du passage de Vénus sur le disque du Soleil, faites à Paris au Palais du Luxembourg, le 6 Juin 1761; avec les déterminations qui en résultent," *Mémoires*, 1761, pp. 81-86.

—————. "Observation du passage de Vénus sur le Soleil, faite à Paris le 3 Juin 1769, dans l'Observatoire du Collège Mazarin," *Mémoires*, 1769, pp. 417-432.

—————. "Observations faites par ordre du Roi, pour la distance de la Lune à la Terre, à l'Observatoire royal de Berlin, en 1751 & 1752," *Mémoires*, 1751, pp. 457-480.

—————. "Observations of Venus on the Sun, made at Paris, June 6, 1761," *Phil. Trans.*, LII, 40 (1761), 216.

—————. "Observation qui prouve que le diamètre apparent de Vénus ne diminue pas sensiblement, lors même qu'il est vû sur le disque lumineux du Soleil," *Mémoires*, 1762, p. 258.

—————. *Du passage de Vénus sur le Soleil qui s'observera en 1769.* Amsterdam: 1769.

—————. "Remarques pour la justification des calculs du passage de Vénus, inséré dans la connaissance des temps de 1761," *Mémoires*, 1761, pp. 107-111.

—————. "Remarques sur le passage de Vénus qui s'observera en 1769," *Mémoires*, 1768, p. 236.

—————. "Remarques sur les différentes observations du passage de Vénus faites en Angleterre," *Mémoires*, 1769, pp. 539-542.

—————. "Remarques sur les observations du passage de Vénus, faites à Copenhague & à Drontheim en Norwêge, par ordre du Roi de Danemarck," *Mémoires*, 1761, pp. 113-114.

—————. "Remarques sur les observations du passage de Vénus, faites à Tobolsk," *Mémoires*, 1761, pp. 111-112.

—————. "Remarques sur les observations faites par M. Pingré à l'isle Rodrigue dans l'océan Ethiopique, pour la parallaxe du Soleil," *Mémoires*, 1761, pp. 93-95.

————— and Messier. "Observations of the Transit of Venus on June 3, 1769, and the Eclipse of the Sun on the following Day, made at Paris, and other Places," *Phil. Trans.*, LIX, 50 (1769), 374-377.

Landen, J. *Animadversions on Dr. Stewart's Computation of the Sun's Distance from the Earth.* London: 1771.

La Roncière, C. de. "Bibliothèques de la Marine," *Catalogue générale des manuscrits des bibliothèques publiques de France.* Paris: 1907.

Leeds, J. "Observation of the Transit of Venus on June 3, 1769," *Phil. Trans.*, LIX, 61 (1769), 444-445.

Lefebvre, B. *Les passages de Vénus sur le disque solaire.* Louvain: 1883.
Of uncertain scholarship.

Lefranc, A. *Histoire du Collège de France depuis ses origines jusqu'à la fin du premier empire.* Paris: 1893.

"Legentil de la Galaisière," *Biographie Universelle* (Michaud), XXIII, 618.

Le Gentil, G.J.H.J.B. "Dissertation sur le diamètre apparent du Soleil," *Mémoires*, 1752, pp. 668-701.

————. "Extrait du journal d'un voyage fait par ordre du Roi dans les mers de l'Inde," *Mémoires*, 1771, pp. 247-280.

————. "Mémoire de M. le Gentil, au sujet de l'observation qu'il va faire, par ordre du Roi, dans les Indes Orientales, du prochain passage de Vénus pardevant le Soleil," *Journal des Sçavans* (Mars 1760), 137-139.

————. "Observations astronomiques faites pour déterminer la longitude de Manille," *Mémoires*, 1768, pp. 237-246.

————. "Observation de la conjonction inférieur de Vénus avec le Soleil, arrivée le 31 Octobre 1751, faite à l'Observatoire royal de Paris; avec des remarques sur les deux conjonctions écliptiques de cette planète avec le Soleil, qui doivent arriver en 1761 & 1769," *Mémoires*, 1753, pp. 43-55.

————. "Observations sur les marées à Madagascar, dans la zone torride," *Mémoires*, 1773, pp. 243-257.

————. "Premier mémoire sur l'Inde, particulièrement sur quelques pointes de l'astronomie des gentils Tamoults; sur Pondichéry & ses environs," *Mémoires*, 1772, II, 169-214; 221-266.

————. "Second mémoire sur l'Inde et en particulier sur Pondichéry et ses environs," *Mémoires*, 1773, pp. 403-436.

————. *Voyage dans les mers de l'Inde* (1760-1771), *fait par ordre du Roi, à l'occasion du passage de Vénus, sur le disque du Soleil, le 6 Juin 1761, & le 3 du même mois 1769.* 2 vols., Paris: 1779-1781.

Le Monnier. "Comparaison des observations du passage de Vénus, faites en Amérique, avec celles qui ont été faites dans le nord de l'Europe," *Mémoires*, 1769, pp. 498-504.

————. "Comparaison du résultat des observations faites sur

la conjonction de Vénus au Soleil, avec le calcul des tables de M. Halley," *Mémoires*, 1761, pp. 192-196.

—————. "Considérations sur le diamètre de Vénus, observé à Tobolsk le 6 Juin 1761," *Mémoires*, 1761, pp. 332-333.

—————. "Contact interne du disque de Vénus au disque du Soleil, observé à Saron par M. le Président Bochart de Saron, & communiqué à l'Académie," *Mémoires*, 1770, p. 232.

—————. "Manière de déterminer l'erreur des tables de Vénus, indépendamment des effets des parallaxes du Soleil & de Vénus, dans l'observation du mois de Juin 1769," *Mémoires*, 1769, pp. 505-508.

—————. "Observation du passage de Vénus sur le disque du Soleil, faite au Chateau de Saint-Hubert en présence du Roi," *Mémoires*, 1761, pp. 72-76.

—————. "Observations du passage de Vénus sur le disque du Soleil, faites au présence du Roi, au château de Saint-Hubert, sous la latitude de 48d 43' 25"," *Mémoires*, 1769, pp. 187-192.

—————. "Remarques sur les observations du passage de Vénus, faites à l'isle Rodrigue le 6 Juin 1761," *Mémoires*, 1761, pp. 88-89.

—————. "Remarques sur un écrit touchant le diamètre de Vénus, qui a été lu à l'Académie," *Mémoires*, 1762, p. 491.

—————. "Suite des remarques sur les observations du passage de Vénus, faites à Rodrigue," *Mémoires*, 1761, pp. 105-106.

Lexell, A. J. "Disquisitio de investiganda parallaxi Solis, ex transitu Veneris per Solem anno 1769," *Novi Commentarii*, XVII (1772), 609-672.

—————. "De parallaxi Solis conclusa ex transitu Veneris per Solem A. 1769 in insula Regis Georgii observato," *Novi Commentarii*, XVI (1771), 586-648.

Libour. "Observation du passage de Vénus sur le disque du Soleil, faite le 6 Juin 1761, au matin," *Mémoires par divers savans*, VI (1774), 435.

Lind, J. "An Account of the late Transit of Venus, observed at Hawkhill, near Edinburgh," *Phil. Trans.*, LIX, 44 (1769), 339-346.

Lindroth, S. (ed.). *Swedish Men of Science, 1750-1950*. Stockholm: 1952.
For material on P. W. Wargentin and T. Bergman.

Littrow, C. L. *P. Hell's Reise nach Wardoe bei Lappland und seine Beobachtung des Venus-Durchganges im Jahre 1769*. Wien: 1835.

Lloyd, H. A. "Description of a Clock by John Shelton, Owned by the Royal Society and Used by Maskelyne on his Visit to St. Helena in 1761, and Probably by Mason and Dixon in Pennsylvania," *Proceedings of the American Philosophical Society*, 94, 3 (June 1950), 268-271.

Lloyd, H. A. "A Link with Captain Cook and H.M.S. Endeavour," *Endeavour*, x, 40 (October 1951), 200-205.

Long, R. *Astronomy*. Cambridge: 1742-1764.
Discusses the 1761 transit.

Loria, G. *Histoire des sciences mathématiques dans l'antiquité Héllénique*. Paris: 1929.

Lowitz, G. M. *Auszug aus den Beobachtungen welche zu Gurief bey Gelegenheit des Durchgangs der Venus vorbey der Sonnenscheibe angestellt worden sind*. St. Petersburg: 1770.

Lownes, A. E. "The 1769 Transit of Venus and its Relation to Early American Astronomy," *Sky and Telescope*, ii, 6 (April 1943), 3-5.

Lubimenko, I. "Un académicien Russe à Paris (d'après ses lettres inédites, 1780-1781)," *Revue d'histoire moderne*, N.S., 20 (Novembre-Décembre 1935), 415-447.
On A. J. Lexell.

————. "La correspondance scientifique de l'Académie des sciences (Outchenaja Korrespondentsia Akademii Naouk)," *Vestnik A.N.S.S.S.R.*, 4 (1934), 27-38.

Ludlam, R. M. "Observations made at Leicaster on the Transit of Venus over the Sun, June 3, 1769," *Phil. Trans.*, LIX, 32 (1769), 236-240.

Luloss, J. "Observation of the same Transit, made at Leyden," *Phil. Trans.*, LII, 48 (1761), 255.

Luskina. *L'Observation astronomique du passage de Vénus par le disque du Soleil qui doit se faire à Varsovie sur la terrace de la bibliothèque publique de Zaluski le 6 Juin 1761*. Warsaw: n.d.

Luynes, C. de. "Observation du passage de Vénus sur le disque du Soleil, faite à Sens, le 6 Juin 1761," *Mémoires*, 1761, pp. 55-71.

Lynn, W. T. "Halley and the Transits of Venus," *The Observatory*, v, 62 (June 1882), 175.

MacPike, E. F. *Correspondence and Papers of Edmond Halley*. Oxford: 1932.

————. *Hevelius, Flamsteed and Halley, Three Contemporary Astronomers and their Mutual Relations*. London: 1937.

Magee, W. "Minutes of the Observation of the Transit of Venus over the Sun, June 6, 1761, taken at Calcutta in Bengal, Latitude 22°30', Longitude East from London nearly 92°," *Phil. Trans.*, LII, 96 (1762), 582.

Maindron, E. *L'Académie des sciences*. Paris: 1888.

Mallet, F. "On the Transit of Venus over the Sun, June 6, 1761," *Phil. Trans.*, LVI, 12 (1766), 72.

Mallet, M. "Extract of a Letter from Mr. Mallet, of Geneva, to Dr. Bevis, F.R.S.," *Phil. Trans.*, LX, 31 (1770), 363-367.

Mannder, E. W. *The Royal Observatory, Greenwich: A Glance at its History and Work.* London: 1900.

Mantoux, P. *The Industrial Revolution in the Eighteenth Century.* London: 1948.

Maraldi. "Observation de la sortie de Vénus du disque du Soleil, faite à l'Observatoire royal le 6 Juin 1761, au matin," *Mémoires,* 1761, pp. 76-77.

————. "Observation de l'entrée totale de Vénus sur le disque du Soleil, faite à l'Observatoire royal le 3 Juin 1769," *Mémoires,* 1769, pp. 245-246.

Marchand, E. *Jérôme Lalande et l'astronomie au XVIIIe siècle.* Bourg: 1909.

Marchand, J. "Le Départ en mission de l'astronome J.-N. Delisle pour la Russie (1721-1726)," *Revue d'histoire diplomatique,* 43, 1 (Octobre-Décembre 1929), 1-26, 373-396.

Marguet, F. *Histoire de la longitude à la mer au XVIIIe siècle.* Paris: 1917.

————. *Histoire générale de la navigation du XVe au XXe siècle.* Paris: 1931.
One of the best histories of navigation.

Martin, B. *Venus in the Sun: Being an Explication of the Rationale of that great Phaenomenon; of the Severall Methods used by Astronomers for Computing the Quantity and the Phases thereof; and of the Manner of applying a Transit of Venus over the Solar Disk, for the Discovery of the Parallax of the Sun; Settling the Theory of that Planet's Motion, and Ascertaining the Dimensions of the Solar System.* London: 1761.

Mascart, J. *La vie et les travaux du chevalier Jean-Charles de Borda (1733-1799), épisodes de la vie scientifique au XVIIIe siècle.* Lyon: 1919.

Maskelyne, N. "Account of the Observations made on the Transit of Venus, June 6, 1761, in the island of St. Helena," *Phil. Trans.,* LII, 36 (1761), 196ff.

————. "Astronomical Observations made at the Island of St. Helena," *Phil. Trans.,* LIV, 57 (1764), 348.

————. "Descriptions of a Method of measuring Differences of Right Ascension and Declination, with Dolland's Micrometer, together with other new Applications of the same," *Phil. Trans.,* LXI, 49 (1771), 536-546.

————. *Instructions Relative to the Observation of the Ensuing Transit of the Planet Venus over the Sun's Disk, on the 3d of June 1769.* London: 1768.

————. "A Letter from Rev. Nevil Maskelyne, B.D.F.R.S. Astronomer Royal, to Rev. William Smith, D.D. Provost of the College of Philadelphia, giving some account of the Hudson's

Bay and other Northern observations of the Transit of Venus, June 3d, 1769," *Transactions of the American Philosophical Society*, I, 111-114.

_____. "Observations of the Transit of Venus and Eclipse of the Sun, June 3d, 1769, made at the Royal Observatory, Greenwich," *Transactions of the American Philosophical Society*, I, 100-108.

_____. "Observations of a Clock of Mr. John Shelton, made at St. Helena," *Phil. Trans.*, LII, 98 (1769), 434.

_____. "Observations on the Tides in the Island of St. Helena," *Phil. Trans.*, LII, 98 (1762), 586.

_____. "A Proposal for discovering the Annual Parallax of Sirius," *Phil. Trans.*, LI, 78 (1760), 889ff.

Mason, C. "Astronomical Observations made at Cavan, near Strabane, in the County of Donegal, Ireland, by Appointment of the Royal Society," *Phil. Trans.*, LX, 39 (1770), 454-496.

_____. "Observations for Proving the Going of Mr. Ellicott's Clock at St. Helena," *Phil. Trans.*, LII, 86 (1762), 534.

_____ and J. Dixon. "Observations made at the Cape of Good Hope," *Phil. Trans.*, LII, 60 (1761), 378-380.

Maury, L. F. A. *Les académies d'autrefois. L'Ancienne académie des sciences*. Paris: 1864.

Mayer, A. *Observationes Veneris Gryphiswaldensis quibus adiecta est M. Lamb. Henr. Röhli Regii observatorii astronomi observationis suae de transitu Veneris per Solem anno 1761 expositio.* Gryphiswaldiae: 1762.

_____. "Observation of the Immersion of Venus on the Sun, June 3, 1769, made at Gryphswald," *Phil. Trans.*, LIX, 39 (1769), 284-285.

Mayer, C. *Ad augustissimam Russiarum omnium Imperatricem Catherina II Alexiewnam, expositio de transitu Veneris ante discum Solis de 23 Maii 1769, jussu illustrissimi & excellentissimi Domini D. Comitis Woldimeri ab Orlow illustr. Academiae Scientiarum directoris suscepta, ubi agitur de fine hujus observationes, 1) cognoscendi veram parallaxin horizontalem Solis, 2) determinandi veram distantium Solis a Tellure, 3) ceterorumque planetarum & commentarum ordinem & distantium, 4) deque commodis inde natis pro geographia, re nautica, physica, &c. adductis ubique observationibus, earumque calculis ac methodus, ipsaque parallaxi hinc deducta.* Petropoli: 1769.

_____. "Expositio utriusque observationes et Veneris et eclipsis Solaris," *Novi Commentarii*, XIII (1768), 541-560; XIV (1769), 111-569.

_____. "On the Transit of Venus," *Phil. Trans.*, LIV, 29 (1764), 161.

Meldrum, A. N. "Lavoisier's Early Work in Science, 1763-1771," *Isis*, XIX, 2, 56 (June 1933), 330-363; XX, 2, 59 (January 1934), 396-425.

Les membres et les correspondants de l'Académie royale des sciences (*1666-1793*). Paris: 1931.

Menshutkin, B. *Russia's Lomonosow.* Princeton: 1952.

Messier, C. "Observation du passage de Vénus au-devant du disque du Soleil le 3 Juin 1769," *Mémoires*, 1771, pp. 501-506.

Mitchell, S. A. "Astronomy During the Early Years of the American Philosophical Society," *Proceedings of the American Philosophical Society*, 86 (September 1942), 13-21.

Mohr, J. M. "The Transits of Venus and Mercury over the Sun's Disk, June 4, and Nov. 10, 1769," *Phil. Trans.*, LXI, 45 (1771), 433-436.

Montucla, J. E. *Histoire de mathématiques dans laquelle on rend compte de leur progrès depuis leur origine jusqu'à nos jours.* Paris: 1799-1802.
Includes material on the transits of Venus.

Mott, F. L. and C. E. Jorgenson. *Benjamin Franklin: Representative Selections.* New York: 1936.

Moulton, F. R. *Astronomy.* New York: 1931.

Mowat, R. B. *England in the Eighteenth Century.* London: 1932.

Murdoch, P. "On the Connection between the Parallaxes of the Sun and the Moon; their Densities; and their Disturbing Forces on the Ocean," *Phil. Trans.*, LVIII, 4 (1768), 24ff.

Nahm, M. C. *Selections from Early Greek Philosophy.* 3rd edn.; New York: 1947.

Nécrologe des hommes célèbres de France, par une société de gens de lettres. Paris: 1770.
A useful collection of *éloges.*

Newcomb, S. "Discussion and Results of Observations on Transits of Mercury, from 1677 to 1881," *Astronomical Papers Prepared for the Use of the American Ephemeris and Nautical Almanac*, I (1882), 363-484.

————. "Discussion of Observations of the Transits of Venus in 1761 and 1769," *United States Nautical Almanac, Astronomical Papers*, II, 5 (1890), 259-405.

————. *The Elements of the Four Inner Planets and the Fundamental Constants of Astronomy.* Washington: 1895.

————. "On Hell's alleged falsification of his observations of the transit of Venus in 1769," *Monthly Notices of the Royal Astronomical Society*, XLIII, 7 (May 1883), 371-381.

————. *Popular Astronomy.* New York: 1878.

————. *The Reminiscences of an Astronomer.* Boston: 1903.

Nordenmark, N. V. E. *Pehr Wilhelm Wargentin, Kungl. Veten-skapsakademiens Sekreterare och Astronom 1749-1783*. Uppsala: 1939.
Contains a French summary.

Olmsted, J. W. "The Scientific Expedition of Jean Richer to Cayenne (1672-1673)," *Isis*, XXXIV (1942), 117-128.

Omont, H. *Lettres de J.-N. Delisle au comte de Maurepas et à l'abbé Bignon sur ses travaux géographiques en Russie (1726-1730)*. Paris: 1919.

Outhier, A. "Autre observation du passage de Vénus, faite à Bayeux, le 6 Juin 1761, avec une lunette de 34 pouces, garnie d'un micrometre dont chaque tours de vis est divisé en 42 parties," *Mémoires par divers savans*, VI (1774), 133-134.

"Du passage de Vénus sur le Soleil; annoncé pour l'année 1761," *Histoire*, 1757, pp. 77-79.

"Du passage de Vénus sur le Soleil, qui s'observera en 1769," *Histoire*, 1757, pp. 99-108.

Pekarski, P. *Histoire de l'Académie impériale des sciences de Petersbourg*. Paris: 1870.
Contains material on Delisle in Russia.

Pigott, N. "On the late Transit of Venus," *Phil. Trans.*, LX, 23 (1770), 257-267.

"Pingré, Alexandre-Gui," *Biographie Universelle* (Michaud) XXXIII, 364-366.

Pingré, A.-G. "Examen critique des observations du passage de Vénus sur le disque du Soleil, le 3 Juin 1769; et des conséquences qu'on peut légitimement en tirer," *Mémoires*, 1770, pp. 558-583.

—————. "A Letter from M. Pingré, of the Royal Academy of Sciences at Paris, to Rev. Mr. Maskelyne, Astronomer Royal, F.R.S.," *Phil. Trans.*, LX, 40 (1770), 497-501.

—————. "Mémoire sur la parallaxe du Soleil, déduite des meilleurs observations de la durée du passage de Vénus sur son disque le 3 Juin 1769," *Mémoires*, 1772, I, 398-420.

—————. *Mémoire sur le choix et l'état des lieux où le passage de Vénus du 3 Juin 1769 pourra être observé avec le plus d'avantage; et principalement sur la position géographique des isles de la mer du sud*. Paris: 1767.

—————. *Mémoire sur les découvertes faites dans la mer du sud avant les derniers voyages des anglois et des françois autour du monde*. Paris: 1778.

—————. "Mémoire sur l'observation du passage de Vénus sur le disque du Soleil, faite à Selenginsk en Sibérie," *Mémoires*, 1764, pp. 339-343.

Pingré, A.-G. "Mémoires sur quelques observations du passage de Vénus, faite le 6 Juin 1761, au-delà de l'équateur; & sur les secours qu'on peut en tirer pour la détermination de la parallaxe du Soleil," *Mémoires,* 1763, pp. 354-357.

——————. "Nouvelle recherche sur la détermination de la parallaxe du Soleil par le passage de Vénus du 6 Juin 1761," *Mémoires,* 1765, pp. 1-34.

——————. "Observations astronomiques pour la détermination de la parallaxe du Soleil, faites en l'Isle Rodrigue," *Mémoires,* 1761, pp. 413-486.

——————. "Observations du passage de Vénus, sur le disque du Soleil faites au Cap François, Isle de St. Domingue, le 3 Juin 1769," *Mémoires,* 1769, pp. 513-528.

——————. "Observation du passage de Vénus sur le disque du Soleil, le 6 Juin 1761, faite à Rodrigue dans la mer des Indes," *Mémoires,* 1761, p. 87.

——————. "Observation of the Transit of Venus over the Sun, June 6, 1761, at the Island of Rodrigues," *Phil. Trans.,* LII, 59 (1761), 371.

——————. "Précis d'un voyage en Amérique, ou essai géographique sur la position de plusieurs isles, & autres lieux de l'Océan Atlantique; accompagné de quelques observations concernant la navigation," *Mémoires,* 1770, pp. 487-513.

——————. "Recherches sur la longitude de plusieurs villes, accompagnées de quelques réflexions sur les nouvelles déterminations de la parallaxe horizontale du Soleil," *Mémoires,* 1766, pp. 17-69.

——————. "A Supplement to Mons. Pingré's Memoir on the Sun's Parallax," *Phil. Trans.,* LIV, 28 (1764), 152.

Planman, A. "An Account of the Observations made on the Transit of Venus over the Sun, June 6, 1761, at Cajaneburg in Sweden," *Phil. Trans.,* LII, 44 (1761), 231.

——————. "A Determination of the Solar Parallax attempted by a peculiar Method, from the Observations of the last Transit of Venus," *Phil. Trans.,* LVIII, 16 (1768), 107-127.

——————. D. D. *Dissertatio de Venere in Sole visa die 6 Junii anni 1761.* Aboae: 1763.

Plummer, H. C. "Jeremiah Horrocks and his *Opera Posthuma,*" *Notes and Records of the Royal Society,* III (1940-1941), 39-52.

Porter, J. "Observations on the same Transit of Venus made at Constantinople," *Phil. Trans.,* LII, 42 (1761), 226.

Précis analytique des travaux de l'Académie des sciences, belles-lettres et arts de Rouen depuis sa fondation en 1744. Rouen: 1814-1821. Contains material on Pingré.

Proctor, R. A.   *Transits of Venus. A Popular Account of Past and Coming Transits from the First Observed by Horrocks A.D. 1639 to the Transit of A.D. 2012.* New York: 1875.
Though technically adequate, it is far from satisfactory historically.

Prony, R. de.   "Notice sur la vie et les ouvrages d'Alexandre-Gui Pingré," *Mémoires de l'Institut national des sciences et arts: Sciences mathématiques et physiques*, I (1796), XXVI-XLVII.

Ptolemy, C.   *The Almagest.* Translated by R. C. Taliaferro (*Great Books of the Western World*, ed. R. M. Hutchins, Vol. 16.) Chicago: 1952.

Quetelet, A.   *Histoire des sciences mathématiques et physiques chez les Belges.* Bruxelles: 1864.
Has some material on Lansberg.

Ramsey, J. F.   "Anglo-French Relations, 1763-1770. A Study of Choiseul's Foreign Policy," *University of California Publications in History*, XVII, 3 (1939).

Rayet, G. and C. André.   *L'Astronomie pratique et les observatoires en Europe et en Amérique, depuis le milieu du XVIIe siècle jusqu'à nos jours.* Paris: 1874-1881.

"Recherche de la parallaxe de Mars et de Vénus, par les observations correspondantes faites au cap de Bonne-espérance & à l'Observatoire de Paris," *Histoire*, 1760, pp. 119-120.

Reid, C. L.   *Commerce and Conquest. The Story of the Honourable East India Company.* London: 1947.

Rigaud, S. P. (ed.).   *Miscellaneous Works and Correspondence of the Rev. James Bradley.* Oxford: 1832.

Rittenhouse, D.   "Calculation of the Transit of Venus over the Sun, as it is to happen June 3d, 1769 in lat. 40°N. long. 5h. West from Greenwich," *Transactions of the American Philosophical Society*, I, 4.

Robinson, H. W.   "Jeremiah Dixon (1733-1799)—A Biographical Note," *Proceedings of the American Philosophical Society*, 94, 3 (June 1950), 272-274.

————.   "A Note on Charles Mason's Ancestry and his Family," *Proceedings of the American Philosophical Society*, 93, 2 (May 1949), 134-136.

Röhl, M. L. H.   *Merkwürdigkeiten von den Durchgängen der Venus.* Greifswald: 1768.

Rose, A.   "Extract of Two Letters from the late Capt. Alexander Rose, of the 52nd Regiment, to Dr. Murdoch, F.R.S.," *Phil. Trans.*, LX, 37 (1770), 444-450.

Rumovsky, S.   "Animadversiones in supplementum cal. *Pingré* ad dissertationem eius de parallaxi Solis," *Novi Commentarii*, XII (1766-1767), 575-586.

Rumovsky, S. *Brevis expositio observationum occasione transitus Veneris per Solem in urbe Selenginsk, anno 1761 institutarum.* Petropoli: 1762.

————. "Expositio observationem occasione transitus Veneris per discum Solis in urbe Selenginsk institutarum," *Novi Commentarii,* xi (1765), 443-486.

————. "Investigatio parallaxeos solis ex observatione transitus Veneris per discum solis Selenginski habita, collata cum observationibus alibi institutis," *Novi Commentarii,* xi (1765), 487-538.

Russell, H. N., R. S. Dugan, and J. Q. Stewart. *Astronomy; A Revision of Young's Manual of Astronomy.* Boston: 1926.

Salluzo, A. *Passagio de Venere sotto il Sole osservato, e calcolato in seminario Romano.* Roma: 1761.

Sarton, G. "Vindication of Father Hell," *Isis,* xxxv, 2 (1944), 97-105.

Schumacher, L. S. *Der Venusstern tritt wieder vor die Sonne den 3 Juni 1769 abends.* Leipzig: 1767.

Shapley, H. and H. E. Howarth. *A Source Book in Astronomy.* New York: 1929.

Shervington, W. "A Letter of Mr. William Shervington to Benjamin Franklin, Esq. concerning the Transit of Mercury over the Sun, on the 6th of May 1753, as observed in the Island of Antigua," *Phil. Trans. Abgd.,* x, 414.

Short, J. "An Account of Mr. Mason's Paper concerning the Going of Mr. Ellicott's Clock, at St. Helena," *Phil. Trans.,* lii, 87 (1762), 540.

————. "The observations of the Internal Contact of Venus with the Sun's Limb, in the late Transit, made in different Places of Europe, compared with the Time of the same Contact observed at the Cape of Good Hope, and the Parallax of the Sun from thence determined," *Phil. Trans.,* lii, 100 (1762), 611ff.

————. "Second Paper concerning the Parallax of the Sun determined from the Observations of the late Transit of Venus; in which the Subject is treated at Length and the Quantity of the Parallax more fully ascertained," *Phil. Trans.,* liii, 97 (1763), 300ff.

————. "The Transit of Venus over the Sun, June 6, 1761, at Savile-House, about 8s of time West of St. Paul's London," *Phil. Trans.,* lii, 33 (1761), 178ff.

Skelton, R. A. (ed.). *Charts and Views Drawn by Cook and His Officers and Reproduced from the Original Manuscripts.* Cambridge: 1955.

Smith, W. "An Account of the terrestrial measurement between the Observatories of Norriton and Philadelphia; with the difference of longitude and latitude thence deduced," *Transactions of the American Philosophical Society*, I, 114-120.

—————. "An Account of the Transit of Venus over the Sun, June 3d, 1769, as observed at Norriton, in Pennsylvania," *Transactions of the American Philosophical Society*, I, 8-39.

—————. "The Sun's Parallax deduced from a Comparison of the Norriton Observations of the Transit of Venus, 1769; with the Greenwich and other European Observations of the same," *Transactions of the American Philosophical Society*, I, 162-180.

————— and J. Ewing, O. Biddle, H. Williamson, T. Combe and D. Rittenhouse. "Apparent time of the Contacts of the limbs of the Sun and Venus; with other circumstances of most note, in the different European observations of the Transit, June 3d, 1769," *Transactions of the American Philosophical Society*, I, 120-126.

————— and J. Lukens, D. Rittenhouse and J. Sellers. "Account of the Transit of Venus over the Sun's Disk, as observed at Norriton, in the County of Philadelphia, and Province of Pennsylvania, June 3, 1769," *Phil. Trans.*, LIX, 41 (1769), 289-326.

Somervogel, C. *Bibliothèque de la Compagnie de Jésus*. Paris: 1898.

Stearns, R. P. "Colonial Fellows of the Royal Society of London, 1661-1788," *Notes and Records of the Royal Society*, VIII, 2 (April 1951), 178-246.

Steward, M. *The Distance of the Sun from the Earth Determined by the Theory of Gravity*. Edinburgh: 1763.

Stone, E. *The Whole Doctrine of Parallaxes Explained and Illustrated by one Arithmetical and Geometrical Construction of the Transits of Venus and Mercury over the Sun*. Oxford: 1768.

Stone, E. J. "A Rediscussion of the Observations of the Transit of Venus," *Monthly Notices of the Astronomical Society of London*, XXVIII (1868), 255; XXIX (1869), 6, 8, 236.

Struve, O. "Lomonosow," *Sky and Telescope*, XIII, 4 (February 1954), 118-120.

"Sur la comparaison du passage de Mercure sur le Soleil arrivé en 1753, avec ceux qui avoient été observés jusqu'alors," *Histoire*, 1756, pp. 90-96.

"Sur la conjonction écliptique de Vénus et du Soleil, du 6 Juin 1761," *Histoire*, 1761, pp. 98-117.

"Sur la conjonction écliptique de Vénus et du Soleil, du 3 Juin 1769," *Histoire*, 1769, pp. 93-101.

"Sur la détermination de la parallaxe du Soleil par le passage de Vénus, du 6 Juin 1761," *Histoire*, 1765, pp. 77-82.

247

"Sur la longitude de plusieurs villes et sur la parallaxe du Soleil," *Histoire*, 1766, pp. 85-89.

"Sur la parallaxe du Soleil, déduite du passage de Vénus, du 6 Juin 1761," *Histoire*, 1770, pp. 74-76.

"Sur la parallaxe du Soleil, déduite du passage de Vénus sur son disque, du 3 Juin 1769," *Histoire*, 1772, pp. 73-77.

"Sur la parallaxe du Soleil, qui résulte de la comparaison des observations simultanées de Mars & de Vénus, faites en l'année 1751, en Europe & au Cap de Bonne-Espérance," *Histoire*, 1760, pp. 108-110.

"Sur le diamètre du Soleil qu'on doit employer dans le calcul des passages de Vénus," *Histoire*, 1770, pp. 79-80.

"Sur le passage de Vénus sur le Soleil, du 3 Juin 1769," *Histoire*, 1770, pp. 80-86.

"Sur le satellite vu ou présumé autour de Vénus," *Histoire*, 1762, pp. 211-219.

"Sur les observations astronomiques, faites pour déterminer la position de Manille," *Histoire*, 1768, pp. 112-116.

"Sur les observations faites par M. l'Abbé Chappe, en Californie," *Histoire*, 1770, pp. 76-79.

"Sur les observations faites pour déterminer la distance de Vénus à la Terre," *Histoire*, 1760, pp. 124-127.

"Sur l'observation du passage de Vénus sur le Soleil, faite à Selenginsk en Sibérie," *Histoire*, 1764, pp. 115-116.

"Sur quelques observations du passage de Vénus, faites au-delà de l'équateur, & sur la parallaxe de Soleil qu'on peut déduire," *Histoire*, 1763, pp. 95-97.

"Sur une erreur qui s'étoit glissée dans les prédictions du passage de Vénus sur le Soleil pour l'année 1761," *Histoire*, 1759, pp. 185-189.

"Sur une nouvelle manière de trouver, avec une très grande précision, le mouvement horaire de Vénus ou de Mercure dans leurs passages sur le Soleil," *Histoire*, 1762, pp. 243-246.

Sutherland, L. S. *The East India Company in Eighteenth-Century Politics*. Oxford: 1952.

Tannery, P. *Pour l'histoire de la science Hellène*. 2nd edn.; Paris: 1930.

Taxil, N. *Oraison funèbre de Pierre Gassendi*. Bordeaux: 1882.

Trébuchet. "Extrait d'une lettre de M. Trébuchet, ancien officier de la Reine, du 18 Décembre 1759," *La Feuille nécessaire, contenant divers details sur les sciences, les lettres & les arts* (24 Décembre 1759), No. 46.

—————. "Extrait d'une lettre de M. Trébuchet d'Auxerre du 3 Juin 1760," *L'Avantcoureur* (9 Juin 1760), No. 21.

Trébuchet. "Lettre au R.R.P.P. journalistes de Trévoux," *Le Censeur hebdomadaire*, III (1760), XXXVI.

—————. "Lettre de M. Trébuchet," *Le Censeur hebdomadaire*, II (1761), LIX.

—————. "Lettre de M. Trébuchet à M. d'Aquin, au sujet du passage de Vénus de 1769," *Le Censeur hebdomadaire*, III (1760), XLV.

—————. "Lettre de M. Trébuchet, ancien officier de la maison de la Reine, a Messieurs les auteurs du Journal des Sçavans, sur l'eclipse du Soleil par Vénus, du 6 Juin 1761," *Journal des Sçavans* (Novembre 1760), pp. 733-737.

—————. "Lettre de M. Trébuchet . . . en reponse a celle d'un Académicien, inserée dans le Journal d'Avril 1761, au sujet des calculs fait par M. Delisle . . . ," *Journal des Sçavans* (Février 1762), pp. 101-109.

—————. "Suite de la lettre de M. Trébuchet," *Le Censeur hebdomadaire*, III (1761), I, 3-14.

Trevelyan, G. M. *A Shortened History of England.* New York: 1944.

Turnbull, H. W. *James Gregory, Tercentenary Memorial Volume.* London: 1939.

"Types and Calculations of the Transit of Venus, and the Eclipse of the Sun and the Moon, in 1769," *Gentleman's Magazine*, XXXVIII (October 1768), 455.

V., F. *Observationes astronomicae anni M. DCC. LXI, in observatorio collegii academici societatis Jesu Tyrnaviae in Hungaria habitae.* Tyrnaviae: 1761.

Van Doren, C. *Benjamin Franklin.* New York: 1941.

Ventenat, E. P. "Notice sur la vie du citoyen Pingré, lue à la séance publique de Lycée des Arts," *Magasin encyclopédique*, VII (1806), 342-356.

Vivielle, J. "Les tribulations d'un astronome dans la mer des Indes," *Communications et mémoires de l'Académie de Marine*, IV, 2 (1925), 4-34.

"Voyage de M. le Gentil," *Histoire*, 1771, pp. 83-86.

Wales, W. "Journal of a Voyage, made by Order of the Royal Society, to Churchill River, on the North-West Coast of Hudson's Bay; of Thirteen Months residence in that Country; and of the Voyage back to England; in the years 1768 and 1769," *Phil. Trans.*, LX, 13 (1770), 100-136.

————— and J. Dymond. "Astronomical Observations made by Order of the Royal Society, at Prince of Wales' Fort, on the North-West Coast of Hudson's Bay," *Phil. Trans.*, LIX, 65 (1769), 467-488.

"Wargentin, Pierre-Guillaume," *Nouvelle biographie générale*, vol. 46, pp. 551-552.

Wargentin, P.-G. "An Account of the Observations made on the same Transit in Sweden," *Phil. Trans.*, LII, 39 (1761), 213.

—————. "Observations of the Transit of Venus over the Sun, June 3, 1769, made in Sweden," *Phil. Trans.*, LII, 42 (1769), 327-332.

—————. "Observations on the same Transit; and on an Eclipse of the Moon, May 8, and of the Sun, June 3, 1761," *Phil. Trans.*, LII, 38 (1761), 208.

—————. "On the late Transit of Venus," *Phil. Trans.*, LIII, 17 (1763), 59.

West, B. *An Account of the Observation of Venus Upon the Sun, the third day of June, 1769, at Providence, in New England.* Providence: 1769.

—————. "An Account of the Transit of Venus over the Sun, June 3, 1769, as observed at Providence, New England," *Transactions of the American Philosophical Society*, I, 91-99.

Wharton, W. J. L. (ed.). *Captain Cook's Journal During His First Voyage Round the World Made in H.M. Bark "Endeavour" 1768-71.* London: 1893.

Whatton, A. B. *The Transit of Venus across the Sun: A Translation of the Celebrated Discourse thereupon, by the Rev. Jeremiah Horrox, Curate of Hoole (1639) near Preston; to which is prefixed A Memoir of his Life and Labours.* London: 1859.

Whiston, W. *The Transits of Venus and Mercury over the Sun at their Ascending and Descending Nodes for Two Centuries and a Half.* London: 1723.

Williams, S. "An Account of the Transit of Venus over the Sun, June 1769, as observed at Newbury, in Massachusetts," *Transactions of the American Philosophical Society*, II, 246-251.

Wilson, A. "Observations of the Transit of Venus over the Sun," *Phil. Trans.*, LIX, 43 (1769), 333-338.

Winthrop, J. "Extract of a Letter from John Winthrop, Esq., F.R.S. Hollisian Professor of Mathematics and Natural Philosophy, at Cambridge, N. England; to B. Franklin, Ll.D., F.R.S., Dated Sept. 6, 1769," *Phil. Trans.*, LX, 30 (1770), 358-362.

—————. "Observations of the Transit of Venus, June 6, 1761, at St. John's, Newfoundland," *Phil. Trans.*, LIV, 49-50 (1764), 279.

—————. "Observations of the Transit of Venus over the Sun, June 3, 1769," *Phil. Trans.*, LIX, 46 (1769), 351-358.

—————. *A Relation of a Voyage from Boston to Newfoundland, for the Observation of the Transit of Venus, June 6, 1761.* Boston: 1761.

Winthrop, J. *Two Lectures on the Parallax and Distance of the Sun, as Deducible from the Transit of Venus.* Boston: 1769.

Wolf, A. A *History of Science, Technology and Philosophy in the Eighteenth Century.* 2nd edn.; London: 1952.

—————. A *History of Science, Technology and Philosophy in the XVIth and XVIIth Centuries.* 2nd edn.; London: 1950.

Wolf, C. *Histoire de l'observatoire de Paris de sa fondation à 1793.* Paris: 1902.

Wollaston, F. "Observations of the Transit of Venus over the Sun, on June 3, 1769; and the Eclipse of the Sun the next Morning; made at East Derehem, in Norfolk," *Phil. Trans.*, LIX, 58 (1769), 407-413.

Wood, G. A. *The Voyage of the Endeavour.* Melbourne: 1944.

Woolf, H. "British Preparations for Observing the Transit of Venus of 1761," *The William and Mary Quarterly*, 3rd Series, XIII, 4 (October 1956), 499-518.

—————. "Eighteenth-Century Observations of the Transits of Venus," *Annals of Science*, IX, 2 (June 1953), 176-190.

—————. "The Solar Parallax and the Growth of International Scientific Cooperation in the 18th Century," *Actes du VIIIe Congrès International d'Histoire des Sciences.* Firenze: 1956. 373-379.

Wright, T. "An Account of the Observation of the Transit of Venus, made at Isle Coudre near Quebec," *Phil. Trans.*, LIX, 37 (1769), 273-280.

Ximenes, A. "Observations of the same Transit made at Madrid," *Phil. Trans.*, LII, 46 (1761), 251.

"Ximenes, Leonardo," *Enciclopedia Italiana*, XXV (1937), 824.

"Ximenes, Leonardo," *Enciclopedia Universal Illustrada, Europeo-Americana*, LXX (1930), 567.

Ximenes, L. *Osservazionne del passagio di Venere sotto il disco scolare, accaduto la matina del di 6 Giugno 1761.* Firenze: 1761.

Zannoni, J. A. R. "Astronomical Observations, made in several Parts of the Kingdom of Naples and Sicily," *Phil. Trans.*, LVIII, 30 (1768), 196.

"Zanotti, Eustachio," *Nouvelle biographie générale*, XLVI, 954.

Zanotti, E. *Veneris ac Solis congressu observatio habita in astronomica specula Bononiensis scientiarum instituti die 5 Junii MDCCLXI.* Bologna: 1761.

—————. "De Veneris ac Solis congressu," *De Bononiensi Scientiarum et Artium Instituto Atque Academia Commentarii*, V, 1 (1767), 126-140; 209-214.

Zinner, E. *Astronomie, Geschichte ihrer Probleme.* München: 1951.

THE DEVELOPMENT OF SCIENCE

An Arno Press Collection

Electro-Magnetism. 1981

Gravitation, Heat, and X-Rays. 1981

Laws of Gases. 1981

Theory of Solutions and Stereo-Chemistry. 1981

The Wave Theory of Light and Spectra. 1981

Ackerknecht, Erwin H. Rudolf Virchow. 1953 *and* Schwalbe, J., editor. Virchow-Bibliographie, 1843-1901. 1901

Airy, George Biddell. Gravitation. 1834

Anderson, David L. The Discovery of the Electron. 1964

Beer, John. The Emergence of the German Dye Industry. 1959

Brown, Theodore. The Mechanical Philosophy and the "Animal Oeconomy." 1981

Candolle, Alphonse de. Histoire des sciences et des savants depuis deux siecles. 1885

Cheyne, Charles H.H. An Elementary Treatise in the Planetary Theory. 1883

Cohen, I. Bernard, editor. Andrew N. Meldrum. 1981

Cohen, I. Bernard, editor. The Conservation of Energy and the Principle of Least Action. 1981

Cohen, I. Bernard, editor. The Leibniz-Clarke Correspondence. 1981

Cohen, I. Bernard, editor. Studies on William Harvey. 1981

Coleman, William, editor Carl Ernst Von Baer on the Study of Man and Nature. 1981

Coleman, William, editor. French Views of German Science. 1981

Coleman, William, editor. Physiological Programmatics of the Nineteenth Century. 1981

Domson, Charles. Nicolas Fatio de Duillier and the Prophets of England. 1981

Donahue, William H. The Dissolution of the Celestial Spheres. 1981

Farrell, Maureen. William Whiston. 1981

Gardner, Walter M., editor. **The British Coal-Tar Industry.** 1915

Godfray, Hugh. **An Elementary Treatise on the Lunar Theory.** 1871

Graetzer, Hans G. and David L. Anderson. **The Discovery of Nuclear Fission.** 1971

Grimaux, Édouard. **Lavoisier: 1743-1794.** 1888

Hall, Diana Long. **Why Do Animals Breathe?** 1981

Hall, Maria Boas. **The Mechanical Philosophy.** 1981

Hannequin, Arthur. **Essai critique sur l'hypothèse des atomes dans la science contemporaine.** 1899

Harvey-Gibson, Robert J. **Outlines of the History of Botany.** 1919.

Heidel, William Arthur. **Hippocratic Medicine.** 1941

Heilbron, John L. **Historical Studies in the Theory of Atomic Structure.** 1981

Helm, Georg. **Die energetik.** 1898

Herschel, J.F.W. **Essays from the Edinburgh and Quarterly Reviews.** 1857

Hiebert, Erwin N. **Historical Roots of the Principle of Conservation of Energy.** 1962

Hilts, Victor L. **Statist and Statistician.** 1981

Hirschfield, John Milton. **The Academie Royale des Sciences (1666-1683).** 1981

Home, Roderick Weir. **The Effluvial Theory of Electricity.** 1981

Kendall, Maurice G. and Alison Doig. **Bibliography of Statistical Literature.** Three volumes. 1962, 1965 and 1968

Maier, Clifford L. **The Role of Spectroscopy in the Acceptance of the Internally Structured Atom, 1860-1920.** 1981

Meyer, Kirstine. **Die entwickelung des temperaturbegriffs im laufe der zeiten.** 1913

Milne-Edwards, Henri. **Introduction à la zoologie générale.** 1853

Morgan, Augustus de. **An Essay on Probabilities.** 1838

Mouy, Paul. **Le développement de la physique cartésienne 1646-1712.** 1934

Olmsted, J.M.D. **Francois Magendie.** 1944

Partington, J.R. and D. McKie. **Historical Studies on the Phlogiston Theory.** 1937, 1938 and 1939

Petit, Gabriel and Maurice Leudet. **Les allemands et la science.** 1916

Priestley, Joseph. **History and Present State of Discoveries Relating to Vision, Light, and Colours.** 1772

Quetelet, M.A. **Letters Addressed to H.R.H. The Grand Duke of Saxe Coburg and Gotha, on the Theory of Probabilities, as Applied to the Moral and Political Sciences.** 1849

Roe, Shirley A., editor. **The Natural Philosophy of Albrecht von Haller.** 1981

Sayili, Aydin. **The Observatory in Islam.** 1960

Schofield, Christine Jones. **Tychonic and Semi-Tychonic World Systems.** 1981

Schweber, S.S., editor. **Aspects of the Life and Thought of Sir John Frederick Herschel.** 1981

Shirley., John W., editor. **A Source Book for the Study of Thomas Harriot.** 1981

Struve, Friedrich George Wilhelm. Études d'astronomie stellaire. 1847

Turner, Dorothy Mabel. **History of Science Teaching in England.** 1927

Woolf, Harry. **The Transits of Venus.** 1959

Wurtz, Adolf. **A History of Chemical Theory, from the Age of Lavoisier to the Present Time.** 1869

Youmans, Edward L., editor. **The Correlation and Conservation of Forces.** 1865

Zloczower, A. **Career Opportunities and the Growth of Scientific Discovery in Nineteenth Century Germany.** 1981